To Dad
with love to you on Fathers D
Nicholas & Caroline.

CW01677424

A Shoot of Your Own

Michael Kemp

A Shoot of Your Own

The amateur gamekeeper at work

Drawings by
Anne Rumsey and Michael Lane R.I.B.A.

Adam and Charles Black · London

First published 1978
A & C Black Ltd
35 Bedford Row, London WC1R 4JH

ISBN 0 7136 1763 2

Kemp, Michael
A shoot of your own.
1. Game and game-birds – England 2. Hunting – England – Management
I. Title
799.2′4 SK311
ISBN 0-7136-1763-2

Filmset and printed in Great Britain by
BAS Printers Limited, Over Wallop, Hampshire

Contents

List of Illustrations

Photographs

Drawings and Diagrams

Some of the material contained in this book has already appeared either in *The Field*, *Shooting Magazine* or in *Shooting Times & Country Magazine*, although I have taken the opportunity presented by publication in its present form to re-write and to expand the original articles.

I am grateful to the Editors for permission to make use of the material in this manner and I am glad to have this opportunity of expressing my thanks.

M.K.

I

The Amateur Gamekeeper Emerges

The winds of change which blew domestic servants, gardeners, grooms and retainers in general out of the lives of all but the rich carried the message that those who wished to survive must learn to fend for themselves. Most of us have come to terms with self-service shops, decorating our own houses, pulling a trolley round a golf course and doing small repairs to our cars. Even the stables, which were once ruled by tyrants who bullied their employers unmercifully, have been invaded by capable little children who clean all the tack, groom the ponies and manoeuvre them in and out of travelling boxes without turning a hair.

Only in the shooting field has the belief that professionals are indispensable lingered on, and that the craft of the gamekeeper is a closed and mysterious business beyond the grasp of those who have devoted less than a lifetime to its study, but this is far from true. I grant that the higher levels demand a certain apprenticeship and that a professional usually works faster, but any untrained amateur who bestirs himself can improve the shooting on almost any piece of land.

How great this improvement may be depends to some extent upon the nature of the ground but the dominant factor is the amount of intelligent work done by the man. Time is necessary; a big stock of game cannot be conjured up in a single year but it will increase steadily provided only that the effort is maintained and the work is done in the one profitable order. This sequence is half the battle. All too often a fortune is spent and all hands are worked to exhaustion, with little to show for their pains, simply because deserving enthusiasts did not know how to set about a straightforward task.

This book describes how to build a shoot from scratch to any chosen level provided that the main objectives are pigeon, duck and pheasants. Grouse and all things which live on heather, as distinct from agricultural land, demand a different technique and partridges are far too difficult for non-experts. Until a novice has won his spurs with pheasants he will be out of his depth with partridges, so his best course is to refrain from shooting them if they are rare.

Pigeon, duck and pheasants, however, give rewards which are out of all proportion to the effort expended. As an example of what can be done, let me tell of a friend who likes using a gun but detests working as a keeper. He rented the shooting of an ornamental lake in an ancestral park, put up some 'No Entry' signs, shot any predators he happened to encounter and scattered one sack of corn to encourage the migrants in the autumn. That was the sum total of the work which was done each year but the resident duck were allowed to breed in peace. Half a dozen guns shot over this place once a month throughout the season and for more than twelve years a blank day was almost unknown. The average bag must have been between 20 and 30 head each day yet this lake, far from being in a wildfowler's Paradise, is only twenty-five miles from central London and in the heart of the commuter belt.

The sequel is worth relating because it demonstrates the vital importance of a few ingredients. Due to a combination of circumstances, the level of the water in the lake was temporarily lowered at the same time as a high proportion of the reeds was removed and the duck were subjected to considerable disturbance. Thus three essentials – peace and quiet, food and congenial surroundings – were removed and the duck vanished overnight. Presumably they took up residence where they were made welcome; at all events they have not yet returned although there is little doubt that good management could persuade them to do so.

The new amateur gamekeepers

Worthwhile shooting has been built from scratch by men of all ages from all walks of life and from most income groups. Some of the oldest are men who could not bear to give up their favourite sport when retirement tightened the purse strings, while at the other extreme are schoolboys who would dearly love to join an established syndicate but cannot afford to do so. Increasing numbers of those who have acquired skill with guns in clay pigeon shooting clubs are discovering the attractions of game and an appreciable number of beaters have set up on their own, while several groups of farmers have taken over from syndicates killed by the recent financial blizzard.

Part of the impetus comes from the fact that there is not enough game to go round and part from the degree by which self-help reduces the cost, but a large proportion is derived from the inherent merits of shooting, both as a field sport and as a recreation apart from guns. It is at last being realised that preserving game takes men away from the artificialities of modern life and puts them where Nature intended, which is in the open air coping with practical problems.

The belief that all shooting men are either wealthy or poachers dies hard. Few spectators appreciate that many an impressive modern shoot owes its foundation and continued existence to an impecunious enthusiast who learned as he went along. With axe and billhook he groomed neglected woods, with unfamiliar carpenter's tools he made every pen, shed and piece of rearing equipment; and the pheasants which now rise in great numbers are the descendants of strays which he welcomed. The cost is exactly what he chooses to spend, and may well be no more than that of playing golf or some similar recreation.

The work is no more exhausting than gardening or golf but it consumes a great deal of time and demands considerable tenacity of purpose. Many willing hands make faster progress but do not affect the final outcome. Some of the better results have been obtained by two friends who held ordinary office jobs and only worked on the shoot in their spare time. Sport to match the finest has been shown by retired men who lived in the erstwhile keeper's cottage and made a full-time occupation of their hobby.

On balance, the best results have probably been achieved by individuals or by very small groups who began on a small scale and expanded. New shoots tend to die young or grow quickly and success seems to attend persistent people who are self-reliant and resourceful. The commonest causes of failure are probably trying to tackle too big an area, enlisting too many members for the shooting available and rearing birds by hand before the beat-keeper's work is up to standard. To these might be added the tendency of willing horses to kick when overloaded.

The first generation of successful shoot builders are always level-headed – they would have failed if they were not – but those who follow sometimes develop delusions of grandeur. If new members forget the down-to-earth pioneers and convince themselves that 'real' shooting consists of standing still on every Saturday of the season while hired retainers drive hundreds of birds over their heads, failure or a change of policy is imminent. That definition of 'real' shooting is a fallacy upon which many sound enterprises have foundered.

The final result is the same whether the work is done by the shooter himself or by a paid employee, but the Do-It-Himself shooting man always parts with much less money and may even make a small profit. Because he knows so much more of the subject, because his perceptions have been sharpened and his palate trained, he derives more pleasure from the actual shooting and from the allied activities such as beating. He also takes up the finest outdoor hobby in the world for those with a taste for such things. From childhood to old age he has an abiding interest throughout the whole of every year; his pleasure is, most emphatically, not confined to the shooting season.

Even so, it is not a hobby to be undertaken lightly. If a man's pleasure in shooting only begins when game is sighted, increases as the bird approaches and his gun swings up, reaches a climax when the brain says 'now' and the bird turns a somersault, but ends when the body thumps upon the ground he will never succeed as an amateur keeper or manager of shooting. If enthusiasm is to be maintained he must enjoy the whole cycle of the shooting year. Fostering the deserving throughout the bleak months of February and March must be a labour of love, helping the game to bring up large families must be his favourite occupation in spring and summer and he must take a savage delight in harrying the rogues who seek to harm them.

The size of the bag at the end of the year will be important to him but the manner of its filling and watching the shoot improve steadily from year to year will give a greater satisfaction. Building up shooting, like doing anything difficult, is interesting by itself. But there is another factor which it shares with all worthwhile sports and games: it is absorbing even when done badly but the fascination grows as skill increases, and there is no limit to improvement.

Finding somewhere to shoot

No-one need confine his shooting to clays if his heart is set upon game, and he should not be daunted by fiscal difficulties. If circumstances allow him to join an existing syndicate he has an easy passage, if he is content with occasional days against pigeon and wildfowl no doubt a few invitations will come his way, but if he wants a place of his own he can have it. Despite all that is said, there is any amount of unoccupied land and renting the shooting rights is chiefly a matter of going about the task sensibly.

Let us consider fair prices first, and the first thing to know is that an equitable rent in one county would be considered extortionate in another. The demand for shooting rights in south-east England is always high and in 1972 25p per acre was paid each year for promising un-keepered land in that part of the country, and double that amount if keepered. Then prices rocketed with as much as £1 an acre being asked for land of low quality, and now they have fallen sharply. Only by repeated enquiries can the state of the market in any particular district be discovered.

Sporting rights can be a major source of income to their owners. It is not always realised that those of Sussex woodlands are more valuable than the timber, but they are not always let, and most certainly not always to the highest bidder. After all, few landowners wish to see a mob of armed strangers on their property and they steer clear of those who might shoot dangerously, allow dogs to chase farm animals, leave gates open or attract unfavourable attention in any way. Men who are well known and have a stake in the community start with an advantage.

Establishing the first foothold is often the most difficult and the most expensive step. After that, a shooter may well be called in by farmers to defend the crops from pigeon and rabbits and his troubles are over when neighbours approach him on the lines of, 'If I let you shoot on my land, will you allow me to join your group?'

The size of the original holding is not of first importance, even one field being better than nothing, but 200 acres of mixed farm land is a good amount for an amateur working alone or with one companion. When that piece is a credit to all concerned, the next target should be whatever area is the minimum for a complete day of driven pheasants. This depends upon the nature of the land, of course, but is about 300 acres in south-east England; and a man of ambition will set his sights upon 1100 acres, because that is a reasonable minimum for a full-time gamekeeper.

What ground to seek

The way in which the land is cultivated sets a limit on the degree to which it can be made attractive to game. Hobson's choice may confine you to an expanse of level ploughland but an area of small fields, thick hedges, numerous woods, ponds, swamps, abandoned clay pits and the like, growing a great variety of crops, will give the best results. Undulating land of this description is the game preserver's Mecca and the more closely your shoot resembles it the better are your prospects.

Two more things should be said. Without the friendly co-operation of the farmer, all is doomed to failure; with it, at least partial success is assured, and much may depend upon whether or not he is invited to shoot. Also, although a wish for a period of mutual probation may cause the initial agreement to be no more than verbal, after the first year a proper lease should be drawn up by a solicitor. There are far too many histories of industrious shooting tenants being deprived of the fruits of their own work, to the landowner's great profit, for any insecure tenure to be acceptable.

Costs and prospects

Every aspirant should know that Count Louis Karolyi built up his shooting until the annual bag was some 10,000 pheasants, 10,000 partridges and 10,000 hares and that the *shooting paid for itself*; that alone should silence defeatists who complain that no poor man is master of his fate in the shooting field. The Karolyi shoots are described more fully in Chapter 13 and are only mentioned here to emphasise that good sport does not necessarily involve great expense.

The rent is a known, fixed charge; after that things cost as much or as little as you care to spend. Admittedly, two tins of poison and a few sacks of corn must be bought but the price of 200 cartridges would cover the essentials; thereafter sweat saves money and brains save sweat. It may even be an advantage to restrict the expenditure to a stated sum: £10 a year plus the rent will swing Nature's balance so far and yield so much shooting. £30 a year would give more, but in the early stages the money can be cut off at any level without losing ground already gained, provided that the work is continued.

True rough shooting, that is to say where no keeper's work is done, is almost without cost but there is little to see except the occasional pigeon and duck. Intelligent work causes the bag to rise sharply, but if it is done by amateurs the value of the game just about equals the additional cost. When rabbits were plentiful (and they seem to be returning) this improved rough shooting often made a small profit and, even without rabbits, it is less expensive than membership of a golf club.

The weakness of improved rough shooting is that no one can depend upon game being present. The best that can be done is to walk-up the open places, work a dog through the thick stuff and then sit in a hide hoping that pigeon or duck will manifest themselves. When they do so in large numbers the shooting may equal anything ever seen on famous grouse moors or expensively stocked estates, but the proportion of resident game is small. The majority are either nomads or migrants in transit, and the guns must seize their opportunities, for birds which are thick on the ground on a Thursday may be sadly absent on the following Saturday.

As long as shooters are content with a few residents plus unpredictable birds in transit, and are their own keepers, the cost need never exceed, say, the price of from 500 to 1000 cartridges each year, in addition to the rent.

To be certain of finding game in any quantity the ground must be keepered and food must be abundant, and that sends the costs soaring. Food is expensive; at the time of writing, a ton of either wheat or barley costs about as much as 1000 cartridges. That amount of grain will, of course, swing Nature's balance a long way; when it is added to the natural food, the land becomes so much more attractive to game than anything available elsewhere that nomads stay longer and residents become more numerous. Basically, however, a ton of grain will feed 600 pheasants for a month and no more. A mallard needs about the same amount of food as a pheasant and a pigeon or partridge only slightly less. Assuming that deserving birds eat all the food intended for them, in theory each ton of corn which is put down raises the average population by something over 50 head.

In an average syndicate which employs one full-time keeper and concentrates on driven pheasants, the cost of the keeper (wages, NHI, cottage, etc.) amounts to about 50% of the total expenditure apart from the rent; and beaters absorb a further 15%. So shooters who fend for themselves halve the costs if they go for driven birds and pay only one third if they walk them up.

Average costs are difficult to learn but the following are the figures from a reasonably well-run driven pheasant shoot for the season 1974–5. The keeper was a part-timer, 1000 pheasants were reared and just under 500 were shot, together with about 500 pigeons, duck and 'various'. The total expenditure was £2000 plus the rent (of which almost exactly £1000 was paid to the keeper) and the value of the pheasants and duck shot was about £400. The value of the other game was ignored. As there were six members, each parted with £266 and, on the average, shot 166 head of game. Had they done the keeper's work themselves they would have had the same bag for £100 per man, plus one sixth of the rent.

The cost of shooting throughout the country is being studied and, although the data is incomplete, the indications are that the figures for the 1974–5 season were much as follows. Confining ourselves to pheasants alone, the highest costs per bird were found in small and medium-sized shoots where there was gross incompetence: figures like £17 per bird were discovered. When all the work was done by an efficient, paid staff the price per bird came out at something between £4 and £7. As this includes the rent and treats all non-pheasants as a cost-free bonus, the result is less alarming than the first impression. Furthermore, extra-good management produced substantially lower figures.

Even so, these lower figures were several times greater than those of some men who were their own keepers, allowed an adequate wild stock to breed in a favourable environment and had some driving but did most of their shooting either while walking-up or sitting in hides. The cheapest shooting of all, and some of the best, was had by retired men who lived on the estate and made a full-time occupation of their hobby; some have made a profit.

It would be a mistake to conclude that the prices in the 1974–5 season are out of date and necessarily lower than those of today. Far from it. The economic climate has caused shooting men to look rather more closely at the accounts, to eliminate waste and to be less ready to remove all difficulties by writing cheques; and the result is that costs have fallen despite inflation. I do not suggest for a moment that bigger bags at lower prices are the be-all and end-all of good management; such is far from the case. But money is one important factor and a cold, hard look at reality ensures that a possible, but not over-modest, target will be selected.

The one rewarding strategy

No matter what level you choose, whether it be slightly improved rough shooting, continuous rapid fire at driven game or something in between, there is only one profitable sequence of work when building up a shoot. It is:

(a) Exclude human trespassers, thus providing peace and quiet for wildlife.

(b) Harry predators so that the deserving have fewer enemies.

(c) Ensure that both resident and visiting game find food.

(d) Provide a habitat, an environment, which game enjoys, and then birds will take up residence and breed. That is what they are longing to do above all else.

(e) Only when these tasks are well and truly done should game be reared by hand; and even then the man will defeat his own purpose if the effort devoted to rearing causes standards to fall elsewhere.

A little reflection shows that to work in any other order is to build on sand. All wildlife benefits from the absence of men and their dogs, and every species tends to assemble in places where it is easy to make a living. If one predator is removed or one sack of corn is added, the stock of game will rise by that much; but more game will eat more food and attract more predators. Only if the food supply is maintained and someone kills the cats, rats and foxes flocking to the easy meals which the treasured birds represent, will the stock of game be increased permanently.

This balance between the stock of game, the amount of food and the number of predators is very delicately poised. Apart from rearing youngsters, all a gamekeeper really does is to tip it in favour of the game. Hard work will move the scales a long way. My friend with the lake did very little but achieved worthwhile results, and each man can choose any amount he prefers, secure in the knowledge that the reward will reflect the effort with precision, provided that the work is done in the one proper order.

2
Starting Work on the Ground

Most shooting leases run from 1 February, so there is still some pigeon shooting after a civil call, without a gun and not at milking time, has been paid on the farmer. A six inches to the mile map with the boundary marked helps conversation and enables you to learn the names of the fields. Then, map in hand, walk all over the new shoot memorising every feature before selecting the best observation point and sitting down for a couple of hours.

The probability is that little game will be seen but the wildlife will bring the news to those who listen. Jays, magpies and pheasants are the gossips of the woods; their alarm notes will locate men, cats and foxes with certainty, but every species has its own danger signals. You will also learn that a surprisingly complete picture of current events can be built up from the sounds alone. A man going about his business in the ordinary way creates an incredible uproar by talking, banging gates, making wires twang and thumping the ground with heavy boots. He draws so much attention to himself that any wild thing which wished to avoid him could do so with ease; and a gamekeeper must do better than that.

Excluding trespassers

Reconnaissance completed, make as many 'Private Land. Please Do Not Enter' signs as are necessary. Two coats of emulsion paint on hardboard with the lettering stencilled will serve. 'Beware of Poisonous Snakes' has proved an effective variation.

Harrying predators

First buy a large tin of Warfarin and another of Cymag, then make either *Game On The Farm*, which is a booklet published by the Game Conservancy, Fordingbridge, Hampshire, or *The Complete Book of Game Conservation* (Barrie & Jenkins) your bedside reading. Both are mines of useful, practical information, refreshingly free from the witch-doctor's hocus-pocus which surrounds so much of traditional keepering.

Friends will be glad to spend a day shooting grey squirrels, crows, jays, magpies, stoats, weasels and domestic cats which are known to have reverted to the wild. Foxes are a special case; the local hunt will draw the coverts on request but if too many survive it is perfectly legitimate to thin them down with gun, rifle, snare or Cymag *provided that the hunt consents*, and most Masters of Foxhounds are reasonable people.

It is counter-productive to quarrel with the local hunt and nothing short of self-destructive to try to be 'clever' with foxes. No matter how artful a man may be there are always dozens who can beat him out of sight. Every veterinary surgeon, most keepers and many countrymen know every trick which either bends the law or smashes it to fragments; a glance often identifies the method, the grapevine twitches and the police start their investigations with the gamekeepers as a matter of routine. Legitimate methods can do all that is necessary, so those who insist on being 'clever' are taking risks without a chance of gain.

Perhaps this is the point where the power of the grapevine in country districts should be emphasised. Very little which is done on farm or shoot can be concealed for any length of time; and the wisest plan is to accept the fact but contrive to have the grapevine working as an ally. The atmosphere will probably be slightly hostile until you become well known.

Tracks in the snow are the easiest way of locating bad characters; failing that, look for footprints in mud, runs in thick growth, droppings and half-eaten bodies. The things to be rid of are rats, feral cats – that is to say, domestic cats which have gone wild – mink, stoats, weasels and grey squirrels. But rats are enemy number one and Warfarin is the weapon to use against them.

First select a target, perhaps the perimeter hedge of a wood, one boundary of the shoot or a pond where wildfowl live. If possible, lay hands on some 6 inch diameter pipes about 2 feet long; the next best thing is to make tunnels from scrap timber 2 feet long with internal dimensions of 7 inches wide and 5 inches high. These tunnels should have a roof and sides but no floor and will be used for traps later. You will also need scores of the 3 inch or 4 inch diameter pipes which are used for subsoil drainage.

Put these pipes at intervals along the selected line, remembering that farmers are proud of the appearance of their land and that the pipes will be

Tracks in the snow. The easiest way to locate predators.

there for years, so camouflage them sensibly. Into each pipe put a pay envelope or twist of cellophane filled with Warfarin and go home knowing that you have done something worthwhile.

Those packets of Warfarin are working for you day and night, ready and able to kill the worst enemy of game, the murderous, thieving, defiling scavenger which spreads all manner of diseases and eats up food intended for its betters, the thrice-accursed rat. The pipes protect the Warfarin from the weather, the envelopes show when the poison is being taken and enable you to remove the baits when you wish.

Carry on methodically in the same way but always choose one target and bait it thoroughly before tackling another. For perfection, do the whole of the boundary first and then work inwards, but this is not vital. Pig-styes and cattle sheds are always haunted by rats; every so often, bait these places heavily for a couple of weeks and then lift the packets and try elsewhere. You will not see many dead rats because they sneak off and die under cover; the good news comes when the baits are not opened.

The effect of Warfarin upon pets and farm animals should be known. A full-grown Labrador which ate half a pound died after six days but I know of no ill-effects resulting from small quantities. Neither do I know of concrete evidence that either cats or dogs can be killed by eating rats or mice poisoned by Warfarin, although there are rumours to the contrary. If the packets are placed where pets cannot reach them, and the tin is kept under lock and key, Warfarin can be used without a qualm.

It would be as well to talk to the farmer about rabbits because some may want to keep them, although most will be glad to be rid of the destructive things. If it is war to the knife, always carry a tin of Cymag with you; then, whenever you suspect that a burrow is inhabited, put a tablespoonful of the powder well inside and close the entrance securely with earth or turf.

On contact with the moisture in the earth or air the powder gives off hydrocyanic gas, which is a deadly and quick-acting poison. As far as the lay observer can tell, death is caused so quickly that this form of gassing must rank as a humane method of control.

A lethal concentration builds up over a length of 2–3 feet in a burrow and lasts, it is said, for twenty-four hours, so if there is only one entrance the death of every living thing within the burrow is virtually certain. But if the burrow has several entrances, it is necessary to find them all and to dose and seal every one. If the holes are scratched open again by rats or rabbits which were outside at the time of the first gassing, it is only necessary to repeat the process.

Notice how simple the control of pests which use burrows has become. The powder is cheap, no skill is required and little time is consumed; but Cymag does demand some sense. It should be kept locked up in a well-ventilated place, not in the house and right away from children. Empty tins should be buried because throwing them into a stream might poison the water. In the hands of responsible people Cymag is safe enough when used in the manner described and the only real danger comes from allowing familiarity to blunt caution. Neither man nor dog should ever sniff the powder and none should be allowed to settle on hands or clothes.

The spooning method which has been described only gives a lethal concentration of gas near the, now sealed, entrance to the burrow; so it is quite possible that undesirables could survive in the depths of a big, complicated system of tunnels. To counter this, pumps have been devised which blow the powder through the entire network. As I understand it, powder is pumped into one entrance until the gas, which shows as a white smoke, comes out of another. Then the smoking hole is blocked and powder is pumped into another until smoke emerges somewhere else; whereupon that hole is sealed and the process is repeated until no smoke escapes.

I am quite ready to believe that this is an excellent method and perfectly safe in trained hands, but I will have nothing to do with it. This is partly because I have yet to encounter a warren which could not be cleared by repeated spoonings, but principally because I fear that I might not notice some white smoke until it was too late.

The Ministry of Agriculture has a pamphlet which sets out their recommendations and regulations governing the use of Cymag and kindred powders. Its title is *Safe handling – Cyanide Gassing Powders.*

With the campaign against rats and rabbits well begun, some attention should be given to crows, jays and magpies. All are confirmed stealers of eggs and eaters of chicks, ducklings and nestlings in general; as such, they are a menace to game in the spring and early summer but do no great harm at other times; nevertheless every one should be killed on sight. I was told by the Director of the Game Conservancy that the shells of 202 game-bird's eggs were found below one crow's nest; so the importance of discouraging these villains can hardly be exaggerated. Rooks, jackdaws and possibly moorhens must be classed as enemies, although their crimes are usually confined to eating eggs.

Every book on preserving game describes a variety of cage-traps, made of wire netting on a frame of some kind, into which all these rogues are supposedly lured by a number of sure-fire baits. I, however, have never seen a single one which paid a dividend. Shooting with a gun over decoys or a bait of a dead rabbit, or with a rifle near their nests, has proved much more effective. As crows and magpies nest before the leaves are thick, the nests are easily found and can sometimes be destroyed; where this is impossible a few shots into them with a rifle at any time when they might contain eggs can do nothing but good.

In the old days, poisoned eggs accounted for great numbers of the crow tribe but the poison is now illegal and there is no effective trap which can be set in or near their nests without breaking the law. Great numbers of egg-stealing birds sometimes assemble in places where a gamekeeper particularly does not want to see them, such as around an un-roofed pen where the breeding stock of pheasants or mallard are supposed to be laying eggs. The rogues will swoop down and take every egg as soon as it is laid if the keeper does not intervene; but this situation should never be allowed to arise because a roof of light, nylon netting excludes all such thieves. In the wild, however, there is no remedy but shooting and the noise of a gun would drive the game away even if the predator were killed. The best solution is to use a ·22 rifle with a hollow pointed bullet, from a distance; and if an orthodox rifle cannot be fired with safety an air-rifle is no mean substitute.

All the larger corvids seem to remember places where they have been badly frightened and to keep away from them. But when making use of this knowledge the keeper must bear in mind that fear affects the number of eggs laid by pheasants. If a gun is fired near a breeding pen, the number of eggs laid in the next twenty-four hours will fall sharply and it will not recover for three days.

Pike and rats are the scourge of wild ducklings. A mallard usually hatches about a dozen youngsters but very few reach maturity, and the same is true of moorhens. Pike cause a lot of the casualties in both cases.

I once knew a pond little bigger than a tennis court, around which mallard habitually nested. For ten years or more there were several broods each spring but not a single duckling survived for a month and it was thought that the water was harmful in some way. Then the pond was drained for cleaning and, to the general amazement, some sizeable pike were discovered although the stream which fed the pond was no more than a trickle. Steps were taken to ensure that no pike could return and the duck have reared their broods successfully on that pond ever since.

Controlling predators in general

Traps and snares deserve a chapter to themselves but the fact of the matter is that Warfarin and Cymag have largely replaced traps, while snares are seldom necessary for anything but foxes.

The best time for attacking all predators, feathered and furred, is the first three months of the year, with March at the top of the list. This is because the cover is never thinner; it is a time of dearth when all wildlife finds it hardest to survive, so a few extra problems will be more than some can solve, and with mating in mind caution is often relaxed. Moreover, it is easier to kill one predator before it breeds than the whole family afterwards. Another effort should be made in May when the naïve young are relatively easy to outwit.

The war against predators never ends because as soon as the resident rogues are destroyed others come across the boundary to fill the void. Perfection can never be attained but Nature's balance is so delicately poised that any reduction in the number of predators results in a big increase in the shootable surplus of game, rather than the total number of game birds. It has been said that a large number of predators simply results in especially alert pheasants with highly developed talents for surviving, but my experience does not support that belief; indeed I am convinced that the predators learn faster than the game and that the incessant harrying of vermin is an essential part of good management. A ready cheque book can flood a shoot with hand-reared birds but it is a poor substitute for genuine skill in the keeper's craft. There must be many, many shoots where £1 devoted to controlling predators would yield better results than does every £20 spent in other ways at present.

Even so, rats, feral cats and mink are the only four-footed things upon which unquestioning war should be waged at all times; not one good word can be said of that trio, but all the others do *some* good. Foxes and stoats kill rats and rabbits, weasels keep mice and voles within bounds and all predators tend to kill the poor specimens of the species upon which they prey: it is the weak, old and stupid which allow themselves to be caught. No wise manager of shooting wishes to exterminate predators but only to maintain a rational balance.

Old-style keepers habitually slaughtered thousands of creatures which are now either regarded as fairly harmless or are protected by law. The modern approach favours peaceful co-existence but reserves the right to eliminate harmful individuals, while the old-timers believed that the whole tribe should be wiped out if one member sinned.

After centuries of persecution, badgers are now protected by law but action may still be taken against the occasional rogue which takes to raiding poultry houses or the like; it is easy enough to snare him at the scene of his crime. Hedgehogs eat the eggs of birds which nest on the ground, so they are not to be encouraged, but there is no sense in trying to eliminate the last one. Moles are harmless from the keeper's point of view except that they sometimes tunnel under the nests of partridges and pheasants, thus causing the nest to collapse and the bird to desert; but that does not justify genocide. There is one exception to this which is that weasels make use of the tunnels dug by moles and have been known to enter pens containing chicks by this route. So if moles are working too close to baby chicks, put some carbide in the tunnel and, with any luck, the mole will dislike the smell and go away.

If a multitude of moles are plaguing the farmer by making hills and tunnels all over his fields, the shooting tenant may earn good marks by thinning them down. The modern technique involves poison but the alternative, which is trapping, is a good starting point for a novice. I dislike killing these resourceful, brave and usually harmless little creatures, but when it is essential it may as well be made interesting.

Some protected predators

All the hawks and owls are protected by law and, on balance, rightly so. Kestrels and barn owls are actively helpful; peregrines, harriers and sparrow hawks are too rare to cause any significant amount of damage, so only little owls and brown owls pose serious threats.

Just occasionally, brown owls run amok and attack half-grown pheasants in release pens; as many as sixty decapitated poults have resulted from one such raid and it is hard to believe that any keeper would permit a second massacre, whatever the law may say. The probability is that the old birds were teaching the youngsters how to kill.

Skills better forgotten

In the old days, poisons which are now illegal played a large part in the control of what we then called vermin. On balance the prohibition is wise, and I am certainly not going to describe how the law can be broken, but as there is a body of opinion which maintains that the pendulum has now swung too far and that keepers are deprived of useful weapons, let us examine the matter impartially.

It is true that foxes and the corvids, that is to say all the crows, magpies, jays and kindred species, are easily poisoned, and it cannot be denied that there is an inescapable element of danger with poisons, just as there is with guns and explosives. The fundamental question is whether the help given by the poison justifies the risk which must be taken.

A trained man, who is careful, can poison foxes with very little risk, but it is essential that he should dispose of the body shortly after death. Should he fail to do so, a chain reaction can start in the following manner. A crow pecks at the fox, carries the poison a short distance, swallows it and dies. A magpie eats the crow, a jay eats the magpie and a cat the jay. How far the trail of death may travel after rats have eaten the cat is a matter for speculation. The same chain reaction starts more easily if poisoned baits are put out for corvids because the birds scatter before dying and are difficult to recover.

That is a terrifying prospect, but the risks are even greater when non-experts lay the baits because, in their ignorance, they use methods which scatter the poison and contaminate the surroundings. Then most wild and domestic animals – horses, cows, sheep, pigs, dogs and anything which eats carrion – are in danger; and that is madness.

To point the moral, there is the story of a new keeper who was appointed to an established shoot. Within ten days he had put down poison, killed four dogs, been detected and sent packing. His successor used legitimate methods, made a success of his job and reigned happily for years.

Transport on the shoot

One of the first difficulties a novice encounters is the problem of moving pipes, sacks of corn, traps, tools and tins of poison all over the shoot without carrying them himself. Land Rovers and tractors can establish dumps but they are too expensive and too noisy for everyday work and they cannot use narrow paths and bridges. The best solution is to obtain a second-hand golf trolley; this is the little, two-wheeled buggy on which bags of golf clubs are trundled round the links. Cut the top off a 5-gallon drum and fix it to the trolley with brackets and straps so that it balances nicely. Then you can pull it for miles with a load of 50 pounds and hardly notice the effort. Pulling one trolley and pushing another doubles the load-carrying capacity.

Avoid the frail models which fold up for transport by car; the club professional will have some robust specimens which will stand up to your work. The first ever made was greeted with derision but after its first day's work I could not lay hands on my own invention until I had made one for every volunteer worker. They are seen at their best when employed on the back-breaking task of carrying drinking water to poults in release pens.

3
Making Game Welcome

Trapping consumes so much time that it is wiser, at this stage, to take the first steps towards making your ground attractive to game rather than to concentrate upon the traps. As soon as trespassers have been excluded, so that peace and quiet reign on your shoot, and predators are even a little less numerous, more game will be seen. This may amount to no more than a few pigeon and an occasional duck or stray pheasant where none came before, but that is enough. If they are made welcome the news will spread and their numbers will grow; better still, they will take up residence if the environment is the slightest bit more attractive than anything which is available elsewhere.

The first step is to ensure that food, water and grit are always obtainable; then facilities for dust-baths, preening and sun-bathing should be added and lastly, because it is a slow process, shelter from the weather, cover from view and places suitable for nesting should be provided. This provision of shelter and cover merges into the subject of 'Habitat', which is discussed in Chapter 6, and the work described in this chapter is no more than First Aid which will suffice until the completion of more urgent tasks allows serious work on the habitat to be started.

Grit

Buy poultry grit at fancy prices if you wish, but the smallest gravel which is used upon the local roads is probably just as good. For perfection give a chalk-based grit, for forming the shells of eggs, and also one with a granite base. Both can be dug – with the owner's permission – from the floors of the appropriate quarries without cost and three or four bucketfuls can be left near each feeding hopper.

Water troughs

When adequate drinking places cannot be contrived in the local streams, water troughs are best made by cutting up old motor car tyres. Saw through the middle of the tread in such a way that two circular rings are produced.

Feeding

Pheasants prefer wheat to barley, pigeon are not choosy and duck take barley before wheat. All of them are fond of acorns and beechmast. Professional keepers insist that daily feeding by hand is the best method, and if it is done properly on every single day they are undoubtedly right. But unless man-power on that lavish scale is available, it is far better to have feeding hoppers as the mainstay and to top up with as much feeding by hand as comes easily.

Hoppers enable one man to look after a far bigger area, less food is taken by the undeserving and they ensure that the game can have a meal even if

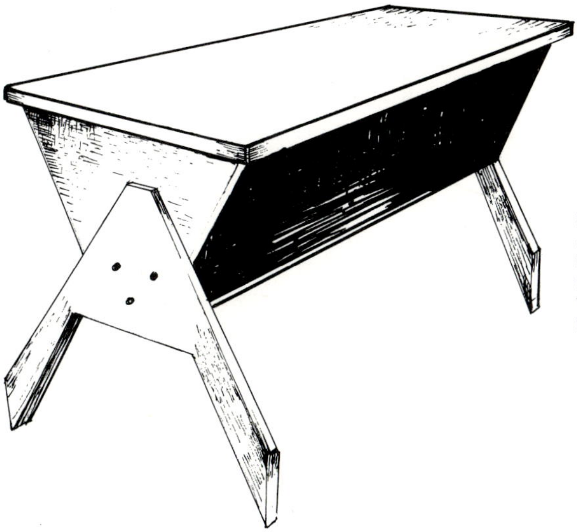

METAL OR PLYWOOD HOPPERS OF ANY SIZE CAN BE MADE TO THIS PATTERN. THE UNDERSIDE IS OF SMALL MESH NETTING THROUGH WHICH A FEW GRAINS FALL WHEN BIRDS PECK UPWARDS.

the keeper has gone abroad. It should never be forgotten that if game find the restaurant closed on three successive days they will go away.

The sketches show a variety of hoppers which can be made at home from empty 5-gallon drums and also a portable bin made of plywood. The latter has wire netting of a small mesh underneath, through which a few grains fall when a bird pecks at it. Drum hoppers can also have mesh underneath but they are more trouble to make and slots have the advantage that their width can easily be adjusted to suit the size of the grain by levering with something like a screwdriver.

Several things about hoppers are important: the grain must never be allowed to become musty because game will not touch it although rats will rejoice. If the corn is not all of one kind, pheasants will pick out the wheat and throw the rest away; and it is essential that all hoppers should be raised to the proper height above the ground. Failure to do so results in feeding squirrels, rats and small birds rather than game. Hoppers should always be suspended from something if cattle or deer can get at them; these creatures soon learn how to overturn any which can be upset and eat the corn.

Notice the tunnel which should be alongside every hopper. Inside it there should be either a trap or a packet of Warfarin to welcome any rat or squirrel which dives into the nearest cover when startled near the hopper. In practice, there are few better places for traps.

I give all drum hoppers two coats of black, bitumastic paint. This is partly for smartness but also because I hope that game will recognise these uniformly black things as sources of food wherever they come upon them.

Feeding hoppers should be placed wherever it is desirable that game should congregate or wherever a feeding point is likely to persuade them not to leave the shoot. Promising places are any clearings or wide rides within a wood, near any flight pond, in a shelter belt, overgrown ravine or similar lengthy strip of cover, or in any patch of rough ground such as an abandoned chalk pit or railway embankment. Excellent short-term results have been obtained by placing hoppers in open spaces within a patch of kale.

It is an advantage if these feeding places are sheltered from the weather and from view and are unlikely to be buried by drifting snow; also they should not be so close to cover that predators can lie concealed and able to pounce. For perfection, hoppers should be sited in pairs with about 50 yards between them so that an aggressive cock pheasant cannot deprive the others of food.

As a very rough guide to the numbers required, it might be wise to aim at one pair in every wood or patch of rough ground with an area of less than 5 acres. In any long, narrow strip of cover, 500 yards between pairs would be about right and in large woods something like 400 yards apart would suffice.

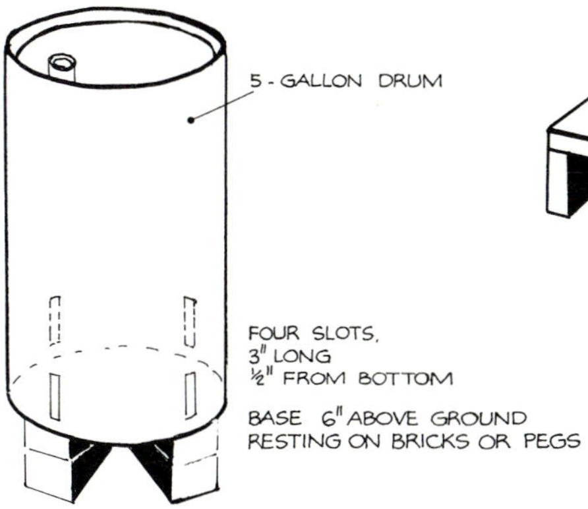

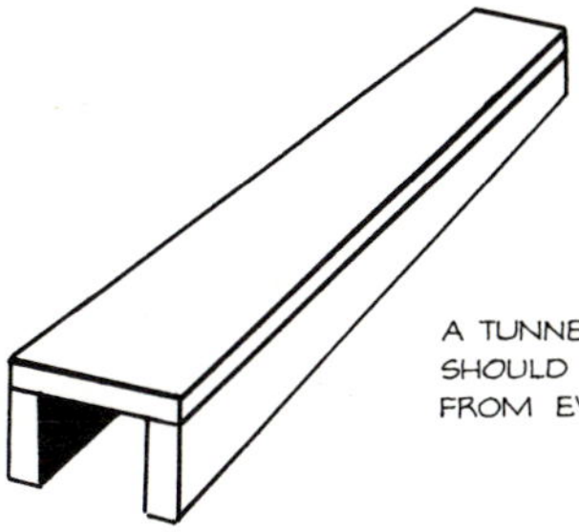

A TUNNEL TRAP SHOULD BE A FEW FEET FROM EVERY HOPPER

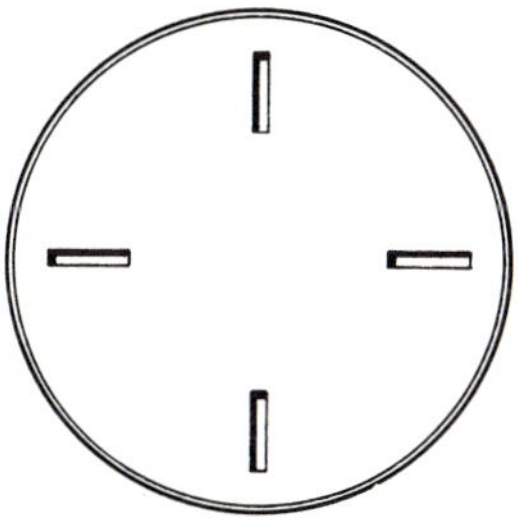

BOTTOM OF DRUM SHOWING SLOTS

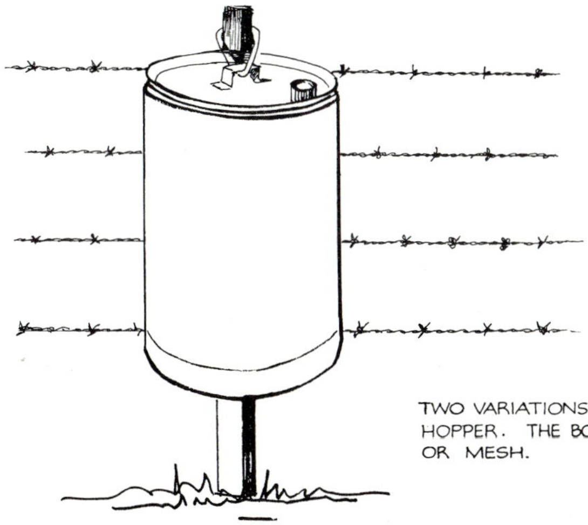

TWO VARIATIONS OF THE 5-GALLON DRUM HOPPER. THE BOTTOM MAY HAVE SLOTS OR MESH.

A device worth considering is the Parsons Automatic Feeder Mark III, which can be obtained from E. Parsons & Sons Ltd., Dept. S/I, Blackfriars Road, Nailsea, Bristol. Essentially it consists of a container for the food, a device for scattering a measured quantity of it, a 12-volt battery to provide the power and a clock to switch it on and off as required. There is also a horn which sounds at feeding time, but this can be disconnected. In these days of inflation prices are changing so fast that it is wiser to say no more than that the cost of this machine is about the same as that of two tons of wheat.

It has a 15-day clock which can be set to feed at stated times as often as is desired in every twenty-four hours; and the amount of corn or pellets in any one feed can be adjusted between the limits of 2 lb and 112 lb. Those I have seen could scatter the food over a circle with a diameter of 25–30 yards.

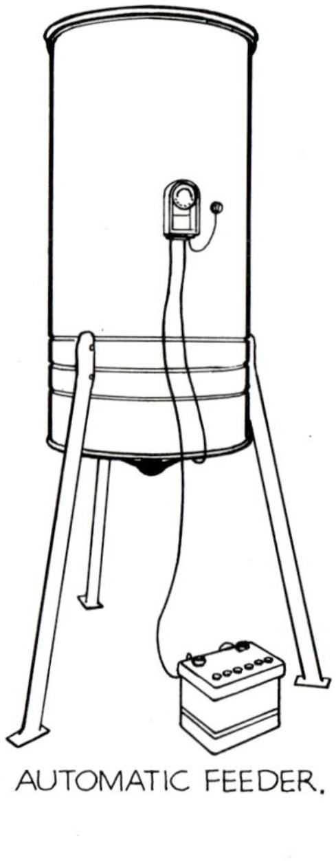

PARSONS AUTOMATIC FEEDER, MARK III

This feeder might be seen at its best near a flight pond; for the correct amount of grain would be scattered without human aid just before the duck were due to arrive. Its powers and limitations are inherent in the design; it has a valuable role, which it plays very well, but it is evident that one feeder would not suffice for a big shoot. The danger from thieves and vandals should also be considered.

How much and when to feed

There is a saying to the effect that a skilled man does more good with one sack of corn than a blunderer can achieve with a ton, but a little thought will go far towards closing the gap. The fundamental thing is that all land, without exception, provides some natural food and the keeper only makes good the deficiencies. In the autumn, when acorns strew the ground, when there are masses of berries and the stubbles are unploughed, there is a great deal of natural food. As this is eaten up, or cannot be reached through a covering of snow, times become harder until spring fills the larder again.

The period of serious hardship is from New Year's Day onwards but a little help from mid-November may not come amiss. When there is ample natural food, nothing is gained by giving more; as supplies run short the hoppers are more in the nature of a bribe to stay on the shoot than a source of essential nourishment; but as February drags into March the food given by man is often all that keeps the game alive.

The commonest mistake is to stop feeding when the shooting season ends, but reflection shows how stupid this is. That is the time, above all others, when game is most in need of help, for without it the birds must scatter or starve. What the shooting man most desires is that large numbers of duck, pigeon and pheasants should assemble on his land able and eager to lay lots and lots of fertile eggs; so why stint the corn which will bring such a great reward? Well fed birds can withstand exposure and resist disease, they will thrive where starvelings will die without a struggle, and hungry parents do not make the best breeding stock.

The first buds do not end the famine and feeding should continue well into the spring. By the time the shooting season ends, all hands may well be sick and tired of toiling on behalf of those wretched pheasants, but a little perseverance is very rewarding.

The folly of luring neighbouring game

The most certain route to failure is to start a vendetta with your neighbours, and the quickest way to do that is to try to lure their birds onto your land. I suggest that it is wiser to make such a success of your shoot by legitimate methods that they wish to amalgamate with you.

It is common knowledge that game, particularly pheasants, can be induced to leave heavily-stocked land for ground which is comparatively empty; and every preserver of game is on his guard against this. Most have known all the tricks since childhood and will trounce a novice in no time if it comes to a duel; but those who have lived with the problem for years are usually fair-minded.

With the exception of rats, all predators are wide-ranging, so by killing some you benefit your neighbour and acquire merit in his eyes. Some of his game will inevitably stray onto your land, survive because you have destroyed the predators which would otherwise have killed it, and may even take up residence with you. As long as this is genuine it should not arouse hostility because, as soon as stability is reached, arrivals will balance departures and each will gain as much as he loses across the boundary.

But to set out with the intention of luring game away is quite another matter. Because the law provides no remedy it is unforgivable, and has been the cause of many life-long feuds in the course of which dogs have been poisoned, stacks set on fire and promises of marriage broken. It is also dishonest.

A newcomer to a country district who is known to be intent on improving a shoot can expect to be approached by a number of shady individuals who will hint that they have secret potions and methods which enable them to charm game from well-stocked coverts onto vacant land as easily as the Pied Piper led the rats and the children; and they are all poor men. Reason insists that if they were as good as they claim they would have become wealthy either as game farmers or as suppliers of game to the wholesale market.

In self-defence I have made a study of their methods since childhood. They range from trickling grain down the inside of the trousers through a hole in a pocket, while ostensibly gathering mushrooms, to scattering corn or raisins which have been treated with one of a variety of high-scented oils. All are effective to a small extent, relatively widely known and therefore quickly detected, and every single one is unprofitable in the long run because they make enemies. The extraordinary thing is that the greener the novice the greater is his faith in such practices, and the more sublime is his confidence that he is too clever to be caught.

First aid for available cover

In these days of big fields and electric fences every scrap of cover is valuable. The first task is to fence every wood and spinney so that farm animals cannot enter. You may have to buy the barbed wire but some landowners will provide the materials if you do the work.

Abandoned railway lines, old clay pits and clumps of gorse or scrub are pearls of great price; so are disused ponds and any rough ground obstructed by rubbish dumps, pill-boxes and the like. With the landowner's consent, run a wire fence round all of them; the work may seem useless in the bleak days of March but the good effects will be apparent once the growth of summer has come. Even the diamond-shaped patches which the plough cannot reach at the base of pylons have value.

The posts for these fences can usually be obtained from the woods, but choose them carefully. In swampy ground, posts of willow often take root, grow and help things along, and there are any number of similar details which repay thought.

Whether any footpaths or bridle-ways which cross the shoot should be fenced along the sides is a matter for judgment. The general rule is that the public must be allowed to pass along such rights of way but may be excluded from the land on either side. In non-partridge country, fencing usually serves little purpose where rights of way cross open fields but it may be worthwhile where they run through woods.

4

Modern Traps and Snares

Many of the old-style keepers were experts with traps, snares and ferrets; that is chiefly what set them apart from men with other skills. For reasons which will emerge, ferrets are no longer essential and those who keep them do so more for their sporting value than from necessity. Snares are seldom used nowadays to catch anything but foxes and the occasional rogue badger and serious trapping is confined to stoats, weasels, mink, grey squirrels and feral cats. A few rats, young rabbits and hedgehogs are destroyed in traps set for other species but they are little more than a bonus. The upshot is that it is neither necessary nor desirable to devote as much time to trapping as was customary in the past.

Nevertheless, trapping and snaring are still essential to good management and no shoot which neglects them can approach its full potential. Nothing I can write will be half as convincing as a study of the tracks in snow. A layman who walks round an average shoot when snow has been lying on the ground for forty-eight hours will probably be astonished by the number of foxes which were moving about, usually heading straight for the places at which game is fed, and the frequency with which the tracks of cats are seen. Any rather unaccountable, small, four-toed tracks were probably made by mink; and at this point you should reflect that each and every one of those foxes, cats and mink is doing an able best to kill something you treasure on at least four days of every week. Follow any trail made by a rabbit which was jinking or running in circles and notice how often it was being hunted by a stoat.

As soon as a man learns that the warnings he has heard about four-footed predators concern him, personally, and have nothing in common with things like typhoons, which cause great distress in faraway places but call for no action in this country, he has passed an important landmark in his shooting education.

New laws, different equipment and myxomatosis, which nearly wiped out the rabbits, have changed the tactics without altering the objective. The main defences against rats and rabbits are now Warfarin and Cymag, which previous generations did not possess. These poisons have removed the need for much of the trapping, snaring and ferreting which consumed a great deal of time in the past; and all the hawks and owls are now protected by law. The highly effective but cruel gin trap has been illegal for years and, although nothing quite so versatile or efficient has yet been developed to replace it, I do not mourn its passing.

The new traps

Both the Mark III and Mark IV Fenn Traps, which can be obtained from A. Fenn, F.H.T. Works, High Street, Astwood Bank, Redditch, Worcs., do well enough against small animals. There are some alternatives, of which the Sawyer and Lloyd and the Imbra traps are examples, but a man who has mastered one can adapt himself to all the others without difficulty.

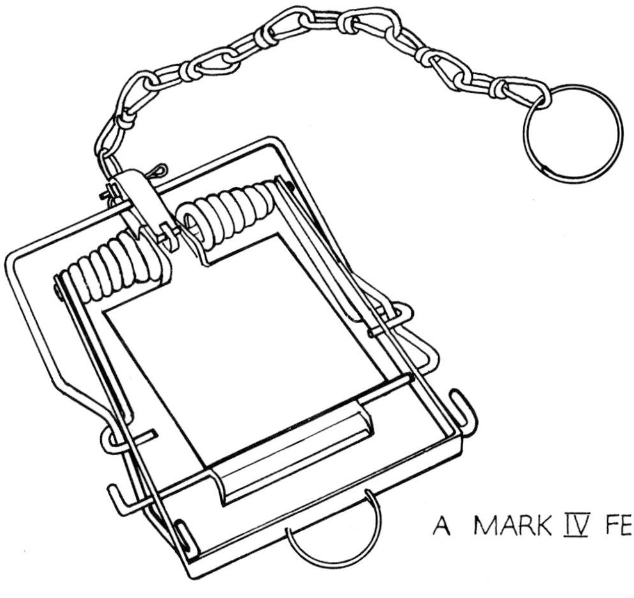

A MARK IV FENN TRAP

What follows is written with the Fenn traps in mind, and although most of the information is applicable to all the makes it should be remembered that some marks of some models require tunnels of a different size.

It must be made unmistakably clear that to set any trap of this kind in the open, on a pole, under water, in a nest or, indeed, anywhere at all except in a tunnel or burrow is to break the law. The only traps which may be set in any other place are box-traps and cages – which are really tunnels in themselves. It must be constantly borne in mind that the anti-field-sports brigade are on the look-out for ammunition and would gladly have our present, sorely-needed traps and snares prohibited.

Reflection shows that with cats as a special case, rats and rabbits controlled by easier means, owls and hawks protected by law and spring-traps forbidden in the open, there are not many potential victims except stoats, weasels, mink and grey squirrels. Fortunately, all of them can be controlled both by Fenns and by box-traps; and the best catcher of cats is a cage which serves very well for mink. This is important because the ability to catch many different species is a feature of any worthwhile trap; nothing is more infuriating than to see a mink and discover that only two of the twenty traps within quick reach will kill a mink and that neither of them will fit into a ready-made tunnel. A variety of traps should be avoided like the plague; everything should catch anything and fit anywhere.

Tunnels for Fenn traps may be either a piece of 6 inch diameter pipe about 2 feet long or a rectangular tunnel of the same length, with two walls and a roof but no floor, whose internal dimensions are 7 inches wide and 5 inches high. I believe that those who prefer 6 inches by 6 inches are mistaken because there is evidence that the victim may be hoisted above the sweep of the jaws if the roof is too high. Tunnels can be made of any scrap timber at least $\frac{3}{4}$ inch thick.

Neither tunnels nor box-traps should ever be treated with creosote because no wild thing will go near them until the smell has worn off. Just leave them exposed to the weather until the scent of the workshop has gone, place them in the selected spots (of which there is more later in this chapter) and camouflage them with earth, turf, logs or whatever is handy. The tunnel should stay put for years; only a novice has to replace and re-camouflage it whenever the trap is re-set.

Leave new traps exposed to the weather for a few days to remove any scent and never lubricate them with motor oil; use something natural which might be found in the woods, such as fat from a pheasant. When setting the trap, gloves are unnecessary but the hands should leave no obviously unnatural smell like petrol. The trap must be well inside the tunnel, beyond the reach of a dog's nose, and the plate must be level with the surface of the earth floor. Drive the peg in first, bury the chain or wire, position the trap and then firm up all the disturbed earth.

A tunnel trap. Notice the sticks across the opening.

The victim must pass over the centre of the plate, so press two sticks into the ground at each end of the tunnel leaving just enough room for a stoat to enter in the middle. Then sweep around both entrances and leave things looking as natural as possible. Whether they have caught or not, all traps should be re-set every week; if this is not done they will become choked by mud and leaves.

Everyone develops their own pet 'wrinkles' while trapping. Some sprinkle a little fine soil on the plate, others use a file until the trap has a hair trigger, but all agree that anything which guides victims towards the tunnel is good. A curved channel dug in the ground, logs laid at an angle, a low wall of turf and pieces of wire netting are all used.

Baits are not essential and should *never* be put on the plate, but anything which makes the neighbourhood of the tunnel worth investigating from a predator's point of view is helpful. A dead rat, hedgehog or piece of meat above the tunnel does no harm; some say that the urine of a bitch stoat acts like a magnet; but if you catch either a female stoat or weasel in May and leave her body in the bushes above the trap you may well kill her half-dozen youngsters within twenty-four hours.

Weasels weigh so little that they may not spring the trap. The best scheme is to skewer something like a pigeon's leg onto the floor of the tunnel; with any luck the weasel will pull like mad and trip the catch.

Box-traps

With all their virtues, and they have many, Fenn traps and those like them have shortcomings. They cannot be made at home and the purchase of fifty or so amounts to a considerable sum. Setting one requires skill, takes several minutes and, above all, is a job which cannot be done well in the dark. Professional keepers tend to prefer Fenns; they are easy to carry, reasonably versatile, a keeper seldom wants to set traps at night and the cost does not empty his pocket.

Amateurs, however, usually choose box-traps because they can be made at home almost without cost, no skill is required to set them and the task can be completed in seconds, even in the dark. Box-traps have always been popular on the Continent, where they are made in all sizes, but they have never been widely adopted in this country, possibly because they are so tiresome to carry from place to place.

The thing not to do with boxes is to make them in different sizes – one for weasels, another for stoats and a third for mink. As we have seen, the ideal trap is versatile and it is foolish to limit the catching capacity of any. Those illustrated, or others just like them, are 30 inches long, were made at home and have caught everything from field mice to mink, including young rabbits and kittens. According to the Game Conservancy a box-trap 40 inches long will catch cats, but I have never tried it. Fault-finders complain that a box only catches from one direction; the answer is to put two boxes side by side, with one entrance facing north and the other south.

The illustrations show the general construction of a box-trap. The case was made from softwood recovered from a demolished building. The width inside is 4 inches, the height is 3½ inches at the entrance and 5 or 6 inches at the closed end. The see-saw is a thin piece of hardwood; both the

(a) A box-trap.

(b) One side has been removed to show the construction. The see-saw is down, awaiting a victim.

(c) The see-saw has tipped up and is now supported by the wire 'U' which is just visible in the centre of the left-hand section.

axle and the U were cut from a wire coat-hanger and one end is covered with two layers of wire netting with a small mesh reinforced with stout wire. A mink can bite through wire netting without this reinforcement.

What the illustrations do not show are the two washers on the axle which prevent the see-saw rubbing against the sides and the balance weights under the see-saw. These can easily be adjusted so accurately that field mice are caught. Making the first trap is a bit of a struggle but it can be done very quickly once you know how to go about the work. I screw the top to the sides, turn it upside-down to fit and balance the see-saw and then screw the bottom on. As long as you use seasoned wood, work reasonably accurately and don't nail it together there is no difficulty.

Once a predator has been trapped, a pinch of Cymag kills it humanely. Tipping it into a stout bag and clubbing is one alternative and submerging the whole trap in water is another. Think well before adopting any other method; a stoat is incredibly quick and bites severely.

There is very little to choose between Fenn and box-traps except their cost and convenience; either can do the work of the other. As both catch mink, I feel that the specialised traps for this one species are a mistake, but some swear by them. An amateur or part-time keeper is probably best equipped with boxes as the mainstay and some Fenns in reserve for emergencies – the tunnels having been built and left without traps. Incidentally, every tunnel without a trap should contain a packet of Warfarin.

Cage-traps

The illustration shows a trigger mechanism which is used all over the world for animals of all sizes, up to full-grown tigers. For mink and cats a frame of thin steel rods covered with any reasonably stout mesh will serve, and the bait for both species is fish. These cages can be made at home without difficulty.

Why predators enter tunnels

It might well be asked why any predator should be so foolish as to blunder into a trap in an unbaited tunnel, but there are two good reasons. Although they are able to go almost anywhere mink, stoats, weasels and rats all tend to hunt along some well-defined line such as a ditch, bank, hedgerow or dry stone wall. When they come to a natural or artificial tunnel they feel compelled to go sniffing and hunting through it; it probably smells of mice and is a convenient route to the other side. When they reach any gap such as

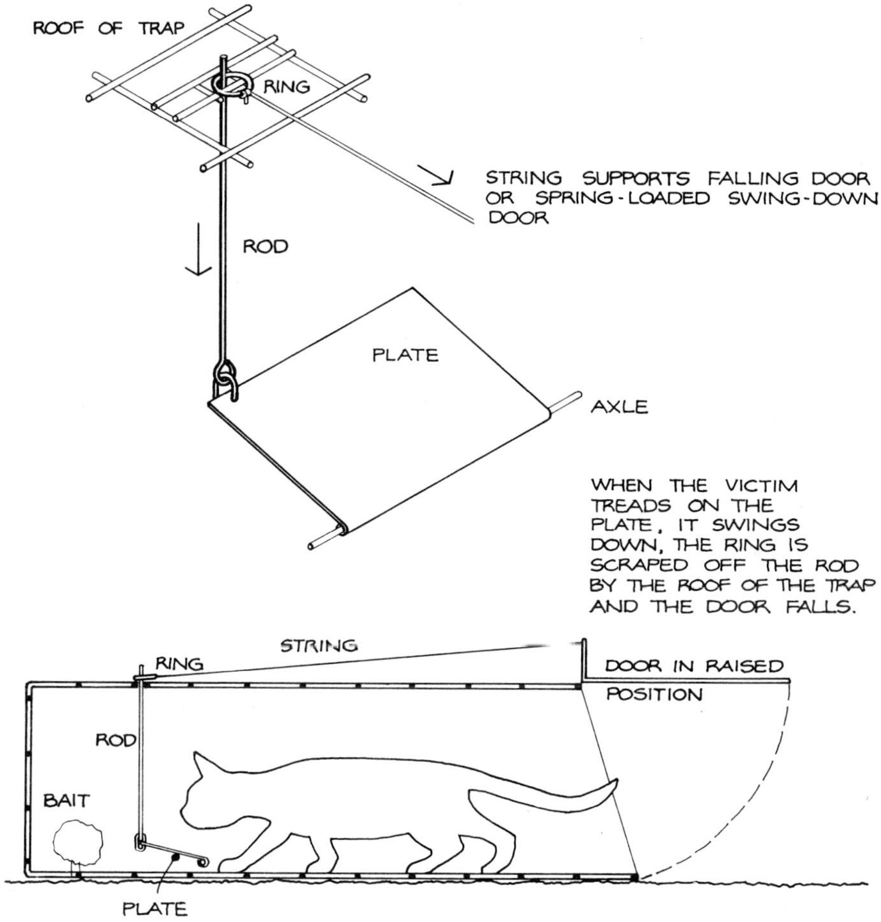

a gate they stop, look round for danger, rush across the open space and dive under cover. If that shelter happens to be a tunnel containing a trap so much the better.

Grey squirrels do not have the hunting motive but they are highly-strung creatures and constantly take cover for no clear reason.

Siting traps

This is one part of the keeper's craft which has remained unchanged for generations, and is shrouded in old wives' tales and hocus-pocus. Some of the lore is based upon a wish to preserve trade secrets or to boost self-importance and some on sheer laziness but, and this is the infuriating thing, a significant part is perfectly genuine. It does not pay either to jettison all the old ways or to follow them blindly. I try to keep an open mind, to respect the traditions until I see a fallacy and then to act as judgment indicates.

Broadly speaking a good trapper is a good naturalist, at least within a limited field, who keeps abreast of current affairs in the world of the wildlife. There is a pattern in the lives of all wild things; although they are able to roam almost anywhere there are places where each species prefers to seek food, to sleep or to relax when there is nothing particular to do, and there are well-defined routes between these places, as well as times when arrival or departure can be expected.

That the pattern is constantly changing as sources of food become available or dry up – with the weather, the phases of the moon and so on – does not alter the fact that it exists; and a good trapper knows what it is and the reason for each change. Indeed, the first hint of trouble is often a disruption of the normal: pheasants flying home instead of walking, many duck visible in the sky at midday and no rabbits feeding at sunset are obvious examples.

Much information comes from tracks, the remains of meals, droppings and broken egg shells, and a method of killing may be typical of a species. For instance, brown owls usually tear the heads off their victims and a dead rabbit with blood on the nape of its neck was almost certainly killed by a stoat or weasel. The behaviour of the wild things is another reliable source of information: pheasants have a distinctive alarm note which announces the sighting of a fox or cat, magpies have a characteristic flight when mobbing a fox, jays are inveterate gossips but their warnings should never be ignored and blackbirds usually have some good reason for sounding the alarm.

By putting all this information together, the trapper builds up a picture of how the local bad characters spend their time, where they hunt, sleep and relax and the routes they are likely to follow; then he places the traps where his enemies are most likely to encounter them. As already explained, predators enter tunnels while either hunting or taking cover, so a good place for a trap would be the junction of any two hunting lines or travelling routes. This might be where two hedges cross or streams meet, the point where the hedge of a field meets the perimeter hedge of a wood, or wherever something long, narrow and thick, such as a shelter belt, offers an attractive route to those who wish to be inconspicuous.

Because no small animal exposes itself to danger in the open if it can avoid it, the pipe which carries a ditch under a gateway is always a good trapping point. And because no animal enjoys swimming in freezing

Box-traps positioned in a wall to catch predators from both directions.

weather, they tend to converge upon any bridge over a stream which is too wide to jump; and there again tunnels may be entered for shelter or in search of prey. It is to be expected that predators will congregate wherever their prey is numerous, so the vicinity of all feeding points should be trapped; but it is less obvious that animals study their own comfort when travelling. This results in the use they make of the outside furrow of a field, rather than pushing through dense, wet grass at the bottom of the hedge, their habit of running along the tops of dry stone walls and of following man-made paths in woods, rather than picking a way through thick, soaking undergrowth.

All healthy wild animals are hardy, so low temperatures do not affect them very much, but they all prefer to sleep or relax in places which are dry, sheltered from the wind and, of course, which give protection from their enemies and good routes for escape. What they try to avoid is any combination of damp, wind and cold. If you let your mind run on these lines, would you not expect that piles of logs within a wood, heaps of large

A tunnel trap set on a bridge across a stream.

stones, abandoned sheds and chalk pits and many hollow trees would find favour with small mammals as sleeping or resting places?

Once a man knows what to look for he can see most of the promising places just as well as the predators can; but there are excellent trapping points which can only be found by trial and error because no human can see the reason for their attraction. And this calls for an explanation. It is well known that there are jockeys for whom almost all horses will put their hearts into their work and go like mad (Scobie Breasley was a notable example) and there are some beautiful riders who have the same effect as a sleeping pill on every horse they mount. I am not referring to bad horsemen with poor 'hands' or faint hearts but to riders who appear highly accomplished to the human eye; no human being can see what they lack yet every horse knows at once. In the same way there is one tree stump of dozens apparently equal upon which fox cubs play every year, one oak in which crows have nested for generations and one minor feature which is a rallying point for stoats or weasels. There must be something special about these places but only experience reveals them to men. By the same token, there are also coverts which pheasants shun for no clear reason, and nothing which man can do will alter that aversion.

Snares for foxes and rogue badgers

I dislike snares but am compelled to use them because there is no other effective legal means of controlling foxes and the occasional rogue badger. The latter are now protected by law and are so interesting that their minor transgressions are readily forgiven, but perhaps once in every twenty years a rogue takes to raiding poultry or pheasant pens and must be destroyed.

Foxes are among the major predators of game. In hunting country the hounds should keep the numbers down and the shooting man should only intervene with the consent of the hunt. The British Field Sports Society's Hunting/Shooting Committee have issued a recommended 'Code of Practice' for the use of snares and it should be noticed that this document contains the sentence 'Masters must take the responsibility for fox control.' In this context 'Masters' means Masters of Foxhounds.

In non-hunting districts a shooting man must fend for himself and snares are the last resort. Only when Cymag in the earths has proved ineffective and when too many have escaped the rifle and the gun should snares be used; but when that time comes this is how to set about the task. Special, heavy duty wires with swivels (and preferably with chains) must first be obtained from a firm which supplies gamekeeper's equipment, such as Gilbertson and Page Ltd., of Hertford. Those which keep the noose tight by means of a no-return slide may dispatch the victim more quickly. The wire used for rabbits or hares is much too frail.

It is essential that the snare should be connected to a secure anchorage by something which is both strong and enduring; preferably by a chain around a tree or fence post. Many a fox has gone off with a noose round its neck after hours of struggling has either pulled the anchor out of the ground or chafed through connections made of galvanised wire, electrical flex or string.

I will not use the so-called 'humane snare' which has a knot to prevent the noose tightening because, although the intention is to avoid killing the wrong animal and to prevent suffering, in my experience the effect is to prolong pain. Once caught, the victim struggles until the trapper arrives, and all chance of instant oblivion from a broken neck or of a relatively quick end by strangulation is denied.

The noose of a fox snare may be circular and 6 inches in diameter but an oval, 8 inches wide and 6 inches deep, is better. The bottom edge of the loop should be the width of your hand, say 4 inches, above the ground in the ordinary way, but experts sometimes increase this to 8 inches. The loop should be held in position by pressing the wire lightly into the ends of two thin sticks which have been split with a knife after they have been driven into the ground.

Snares for badgers are exactly the same as those for foxes except that the bottom of the loop should touch the ground because a badger carries its head very low.

The art of setting snares lies in knowing where to put the snare and non-experts should set them only in hedges, fences or holes in walls. Experts can catch foxes on paths but that calls for a degree of skill which no novice can possess. In practice, you see some holes in a hedge with runs leading to them; and you know that a fox goes in that direction because you have seen its tracks; but you are not certain of the difference between a hare's run and that made by a fox. My solution is to set snares in all the holes because I strongly suspect that hares and foxes often follow the same routes.

Before you set a snare let your mind run on the following lines. Deer have been caught by the leg in snares, so have horses, cows and sheep; dogs have been caught by the neck and men by the ankle. What is likely to happen to you and your shoot if any farm animal, dog or foxhound is killed or injured by a snare you set? Laming a potential Derby winner might prove expensive. The importance of setting snares only where such disasters cannot occur will then be crystal clear.

Every snare must be visited every day; run the loops up if you cannot do so, but inspections at dawn and sunset are better and a gun must always be carried. You may find a dead fox in the open but a fearful tangle of grass, brambles, kinked wire and fox is more usual, and the gun may be needed to end the struggle.

There is one more point. During the pheasant rearing season a few snares are often set around the rearing field as part of the fixed defences, but this is an exception. In the ordinary way they are not used against foxes in general but to catch one particular sinner whose ways are known. Always keep some snares available. The fox which took a nesting duck from the pond will be back for more within forty-eight hours: snares in the shed may provide a fitting welcome but those in some shop will be too late.

Some practical difficulties

The great snag about traps and snares is that they consume a fearful amount of time. Both the law and common humanity insist that every trap should be visited every day, so week-end keepers can do no more than set them on Friday and render them harmless late on Sunday. Moreover, sloshing round the shoot in driving sleet and leaking gumboots to find that the traps have caught nothing is a poor recreation.

A full-time keeper looking after some 1100 acres or more will have about seventy traps set at all times except the height of the rearing season; and every one of them is working for twenty-four hours of every day and seven days a week. Once the predators have been thinned down, they catch very little but their great value is that they stifle trouble before it becomes serious. A by-product is the amount of news which is gathered from the wildlife and from farm workers, postmen and so on while touring the traps. The first inkling of the whereabouts of pigeon, of the habitual route of a fox, of motorists given to shooting from their cars and of ostensible courting couples who wander into the woods at dusk with a torch and a rifle, often comes through the grapevine.

Farm workers may be prepared to look after a few traps in exchange for cash or pigeon shooting, a boy is often delighted to be placed in charge of a line of traps sited along his route to school, and a man who travels to work by car can see into tunnels without stopping. These are all expedients and far from ideal, but if they suffice to keep even a dozen traps going on 250 acres they make a world of difference.

The amount of trapping which is done is always subject to striking a rational balance between the desirable, the possible and the profitable. A fox or a cat is always worth a major effort, but after that priorities must be considered and feeding the pheasants is more important than trapping squirrels; even so, the following story emphasises that wishful thinking should not decide the issue. A smallholder lived between a vast area of heathland to which the public has access and a shoot which reared pheasants on a large scale without bothering much about predators. Exasperated by the damage done by foxes he set about them; and whether he killed 180 in two months or 280 in three months (both totals were rumoured) does not matter. The point is that foxes were so thick on the ground that neither game nor any other wildlife could prosper, and a counter-attack was long overdue.

Two questions spring to mind concerning this massacre. The first is, 'What did so many foxes find to eat?' And the answer is, 'The debris left by picnic parties on the heath, hand-reared pheasants and the contents of dust bins.' Whether this accounts for the fact that a large proportion had mange I do not know. The second question is, 'What in the world was the local hunt doing?' and it is unjust to reply, 'What indeed?' because that smallholding stands where the boundaries of three hunting countries meet, so perhaps all were handicapped.

Outmoded traps and snares

Almost any animal from an elephant to a rat, together with many birds, can be snared; it is only a question of setting a suitable noose in the right place. Also, most snares have a variation by which the victim is hoisted into the air as a bent sapling straightens or a heavy weight falls. All of them can be interesting but they are out-moded in the sense that their work can be done better by other means.

In the same way, there is a vast variety of traps which can be made from local materials by anyone who has an axe and a sharp knife; but none of them serve an English gamekeeper's purpose as well as the box-traps and Fenns which have been described. There might be one possible exception to this because an ordinary hazel-rod pheasant catcher can be used for trapping grey squirrels if covered with suitable wire netting; and with or without netting the same catcher is ideal for removing pigeons which make a nuisance of themselves in suburban gardens.

Ferrets

The role of ferrets has changed. In the past they were essential because there was no other way of coping with rats or rabbits in burrows. A complete book could be written about their breeding, housing, feeding and handling; but their work can be done better, in a fraction of the time, by an unskilled man with a tin of Cymag.

Ferrets are still kept for their sporting value; their points are discussed in detail by the knowledgeable, while the whole paraphernalia of lines, collars, muzzles, carrying boxes, nets, little bells and special spades is made ready. On a good day all hands enjoy themselves and put some rabbits into the larder, but on a bad one no rabbits bolt and the ferrets stay underground while men dig through tangled roots to recover them – and often fail to do so. On a really bad day, and they are not as rare as might be supposed, everybody is bitten at least once and someone is shot; for shooting bolting rabbits safely calls for very cool heads. Nevertheless there are men who enjoy the sport; one of them was T. N. Sedgwick and there is a chapter on ferrets in his book *The Young Shot*, published by A. & C. Black.

5

Strategy and Methods Reviewed

At this stage it might be wise to stand back from the details and reflect upon the general plan. In the first few months little but pigeon shooting can be expected but as the work which has been described in the previous chapters takes effect, there will be more game on the ground. The main objectives are pigeon, duck and pheasants, and the wildfowl may well be the first to respond to help. But hares, snipe and woodcock will be seen more frequently and any partridges present will be less likely to vanish.

If work is continued on the same lines as before and a start is made upon the habitat, the progress will continue until a certain level is reached, and this will be different on every shoot. But it will be slow and there is a temptation to plunge into rearing pheasants or duck in bulk. For a number of reasons this is seldom successful, but it is probably the most common mistake and it is certainly the most expensive. To release hand-reared pheasants onto poorly-keepered land where the habitat leaves much to be desired is as unrewarding as trying to fill a bath without putting in the plug. It is impossible to give accurate figures but it is long odds that 90% of the young pheasants will be dead within a month of leaving the release pens, and that the survivors will hasten to more congenial surroundings as soon as they can find them.

I agree that in a minority of cases there is no satisfactory alternative to concentrating upon reared birds; but it is almost always wiser to foster the game breeding in the wild and to improve the habitat until the whole shoot is in apple-pie order before rearing on any but the smallest scale. Even then, a most careful balance should be struck because it may be more profitable to devote the cash and effort to enlarging the shoot rather than to rearing by hand.

My suggestion is that perfection would be reached if the wild birds provided an acceptable amount of sport and rearing was unnecessary. This can seldom be achieved in modern conditions in this country and the next best course is to aim at a high standard of beat-keeping and habitat but to supplement the wild game by rearing in one's spare time. The vital thing is that the effort devoted to rearing must not cause standards to fall elsewhere.

When this plan is adopted, the land becomes more attractive to game each year; more birds breed in the wild, a higher proportion of the young survive to maturity and smaller numbers are lost through natural wastage. This term, 'natural wastage', is part of the game preserver's jargon and means 'game which is lost through causes other than shooting'. In much the same way 'beat-keeping' means all the gamekeeper's work other than rearing by hand, and it may include work upon the habitat.

A millionaire with unlimited man-power available could press on with the beat-keeping while rearing large quantities of game, but it is worth considering what happens when concentration upon rearing causes the beat-keeping to be neglected. As soon as the keeper withdraws to the rearing field the predators, hitherto prowling just outside the boundary, move in, like what they find and summon their friends. They give the mature birds nesting in the wild a rough passage and they slaughter the chicks and ducklings; but the four-footed predator's idea of Paradise is a stream of hand-reared pheasants, about eleven weeks old, fluttering out of release pens. A pampered Jack Russell can catch them, so foxes, cats, stoats and mink make merry. Worse still, the greater the number of poults which reach the woods as a result of Herculean labours in the rearing field the more do the predators rally to the pickings. And when the shooting season comes, such misguided men crown their folly by saying, 'This is not as good as we hoped; let's ignore the expense and rear twice as many birds next year.' If they would only apply the effort where it does the most good they would have far more shooting for, literally, a fraction of the money.

The final result of sacrificing the beat-keeping to rearing is that predators multiply and the habitat and the wild game decline to such an extent that there is nothing to shoot except a small proportion of the hand-reared birds. There is no progress from year to year; every spring you have to start all over again; and that is how to produce the most expensive birds of all.

The theory of the shootable surplus

Have you ever wondered why the amount of game on the ground does not vary very much from year to year whether it is shot or not? A few years ago an outbreak of foot-and-mouth disease restricted shooting and the general opinion was that a great number of pheasants would survive to breed in the woods so that the following season would be outstanding; but it was not. In the same way, hundreds of thousands of pigeon were shot in the hard winter of 1962–3, yet twelve months later they were as numerous as before. An example from a longer period of time is that the numbers of pigeon, duck, partridges and pheasants did not change greatly throughout the Second World War when there was little shooting for sport.

The most credible explanation is that shooting on a reasonable scale does not control the number which survive the winter to breed; that had there been no shooting there would have been equal losses from starvation, exposure, predators, disease and so on. Let us imagine that fifty pheasants are confined in a pen but they are given only enough food for thirty each day. The greatest number which could survive to breed is thirty, and this number would not be affected if twenty were shot before they starved, but the breeding stock might be reduced if more than twenty were shot.

This example is over-simplified, of course, but it illustrates the general idea. Broadly speaking, the way to increase the stock of game is to shoot within reason and to minimise natural wastage while giving the survivors every opportunity of rearing large families. After all, losses from starvation, exposure, disease and straying can all be diminished by man, as can the toll taken by predators; and increasing the number of youngsters which reach maturity is largely a matter of allowing the parents to get on with what they want to do in favourable surroundings. Those three things – avoiding losses, removing enemies and providing congenial conditions – are the essence of a beat-keeper's work because they increase the shootable surplus.

The final decision on strategy

With certain reservations, my vote would always be cast in favour of the following policy. If the shoot were no bigger than 250 acres I should work on the beat-keeping until it was beyond reproach and I should start on improving the habitat. My objective would be to foster the wild game so that rearing would not be necessary; I should not expect to achieve entire success but I should work in that direction.

With the heavy work on the habitat completed (that is to say when the shrubs, coppice and young trees were planted but still too small to do their work) I should think about rearing a few birds but should probably enlarge the shoot to 350 acres instead. When the beat-keeping and work on the habitat throughout that 350 acres was well in hand I should probably rear as many birds as I could without prejudice to the work elsewhere; this might be from 200–400 pheasants. And when that was done I should extend the shoot, in stages, to 1100 acres, making certain that the extra work did not cause standards to fall in the older areas.

At all times my primary concern would be the game breeding in the wild. I should not expect them to provide an adequate shootable surplus in the early stages, but I am reasonably certain that there would be steady progress towards that goal from year to year. In theory, at any rate, rearing would always be a supplement rather than the mainstay.

If it is thought that such a programme is unduly cautious, reflect that you can run 250 acres single-handed, the financial liability is small, it is an excellent training ground where the mistakes which you will inevitably make will not amount to disasters and you will learn both how much things cost and how long every job takes. That knowledge is extremely valuable when the time comes to recruit allies or engage staff.

I said that the foregoing plan would be my choice subject to certain reservations and I should be willing to start on a much more ambitious scale if ample funds and willing workers were available from absolutely reliable sources. Nothing would induce me to start on, say, an expensive rearing programme unless the cash for that year was in the bank and it was long odds that funds for the two following years would be available. Even a charming letter in mid-September is pretty devastating when it says that the writer's subscription will not be forthcoming and you are faced with the choice of putting up the money yourself or cutting the losses, turning the birds loose to fend for themselves and wrecking the shooting season; so I do not allow the situation to arise.

In the same way I should hesitate to employ a man I did not know well if there was a danger that he might become indispensable. In this connection I am influenced by the story of an employee who was in charge of a rearing field and chose the moment when 1500 chicks were actually hatching to announce that he would withdraw his services, there and then, unless certain conditions were met. Had the employer not been ready and able to act as understudy and replace the professional he would have been in an impossible position because, unless he gave way, the 1500 chicks would have died and the shooting season would have been ruined.

Work in spring and summer

During the first year on a new shoot there will probably be no slackening of effort during the nesting season. Excluding trespassers, attacking predators, establishing feeding points and making a start upon improving the habitat are the top-priority tasks; and nothing which will endure can be achieved until a reasonable standard has been reached with those basic tasks. The following suggestions may, therefore, apply more to subsequent years. There is always a tug-of-war between the interests of the forester and those of the gamekeeper, even when the two are one and the same man, so the best that can be done is to find a sensible compromise.

When the nesting season approaches, the ground should be disturbed as little as possible. What you most desire is that lots and lots of pigeon, duck and pheasants should prefer your land to all others for nesting. All they ask is peace and quiet, food and shelter; if you provide them, the birds will nest. How successfully they bring up their families largely depends upon how good you are at killing their enemies without disturbing the nursery.

Warfarin and Cymag work in silence without much human assistance and are, therefore, almost ideal. To fire a gun at a grey squirrel in May is sheer stupidity because the noise does more harm than the target; if an orthodox ·22 rifle cannot be used safely an air-rifle is no mean substitute.

As we have seen, the major pests are foxes, cats and crows; an entire day devoted to killing any one of that trio would be well spent. The lesser sinners are jays and magpies. Traps are the main defence against stoats, weasels and mink but any you happen to meet should be shot.

The keeper wants to maintain a cloistered calm in all the woods throughout the spring and summer but that is just when the forester has the best opportunity of doing his cutting, thinning and planting. The only time when they are not in opposition is during February and the first half of March, so a rational balance is the only solution and this must be accepted when the programme of work for the year is drawn up.

Two more things should be kept in mind: your very presence disturbs the game you are anxious to protect, so you make a loss if you visit a place without doing something useful; and all predators are good at self-preservation. You will kill very few unless you hunt them intelligently. Despite the maxim that a keeper should never be without a gun, it seldom pays to mix serious work with killing vermin; it is far more effective to finish the job and then to concentrate on the predator.

6
Improving the Habitat

A dictionary defines 'habitat' as 'the natural home of a plant or animal', but in connection with game it has much the same meaning as 'environment'. The nature of the terrain is implied together with the trees, crops and vegetation in so far as they affect the well-being of any particular species. To improve the habitat of the pheasants is to make life easier for them by changing the physical conditions in which they live, perhaps by providing places suitable for nesting or by planting Christmas trees to shelter their roosting places from cold winds.

It is true that an environment ideally suited to one species might ensure the extinction of another, but in the narrow context of lowland, English game any change which benefits one species will be helpful to most, albeit to different degrees. To the extent that no building is better than its foundations, it can be said that no shoot is better than the habitat; but the same is true of freedom from predators, disease and disturbance, together with the amount of food available.

If you have the good fortune to shoot over land which has small fields, wide hedges, patches of rough ground, reed beds, abandoned gravel pits, nice warm woods and a great variety of crops, trees, bushes and weeds, give thanks to your fairy godmother. If not, just work steadily to provide as many of them as you can with the landowner's consent.

In the course of this chapter I hope to make clear that variety itself is a virtue; that most features affect more than one issue, and that almost no one place is ideally suited to any species throughout the year. A source of food, for instance, may also provide shelter from the weather and cover from predators; while there must be something to replace the richness of the cornfields after the stubbles have been ploughed if game is to remain on the property. A group of oak trees may well provide cover in the summer and food in the autumn, but it offers little to game when the urgent need is for shelter from north-easterly gales in February, so some other good shelter must be available then if the game is to prosper.

Most agricultural land has less cover than a preserver of game would wish, and that is why the work described in the section 'First aid for available cover' on page 24 should be given a high priority. The remainder of the improvement of the habitat is better left until more pressing tasks have been completed. For clarity it is convenient to think of short-term improvements, long-term and those designed to present the game at its best, but in practice most of the work can be done at the same time.

Short-term improvements

The man who can make the place extremely attractive to game in a single year is the farmer; but he knows full well that to farm for the shooting is to go broke. It is true that a doting millionaire could have superlative sport very quickly if he left the stubbles unploughed till springtime and arranged the crops so that the land looked like a chess board, with small patches of roots, kale, stubble, rape, clover and potatoes divided by thick hedges and nice, weedy banks. The shooting might be excellent but the farm would certainly lose a great deal of money and no serious farmer will do more than plant the crops which he wants in any case in places beneficial to the shoot; but that alone can be a tremendous help.

Let's see how this can be done and let it be assumed that the farmer intends to plant 5 acres of kale for his cattle. I have never understood why kale, which is always the wettest place on the farm, should attract game, but it does. If these 5 acres were planted in one big block 160 yards long and 150 wide, surrounded by thick hedges, it would be difficult to flush the birds because they could run about unseen and steal away along the hedges. But if the 5 acres were made up of a number of long, narrow strips, a few beaters could easily flush everything.

Both kale and roots are sometimes planted alongside a wood to increase its effective size after the corn has been cut; but apart from this, the shooting man likes to see these crops on the tops of ridges and set well back from the places to which the birds will fly when they rise. A farmer who is keen on shooting will often co-operate on these lines, and will always arrange his work so that the kale is disturbed the day after the shoot rather than the day before. If relations are less than cordial, the shooting tenant may find that tractors, dogs or something else raise Cain in the kale on every shooting morning.

It was once a common practice to plant alternate, narrow strips of kale

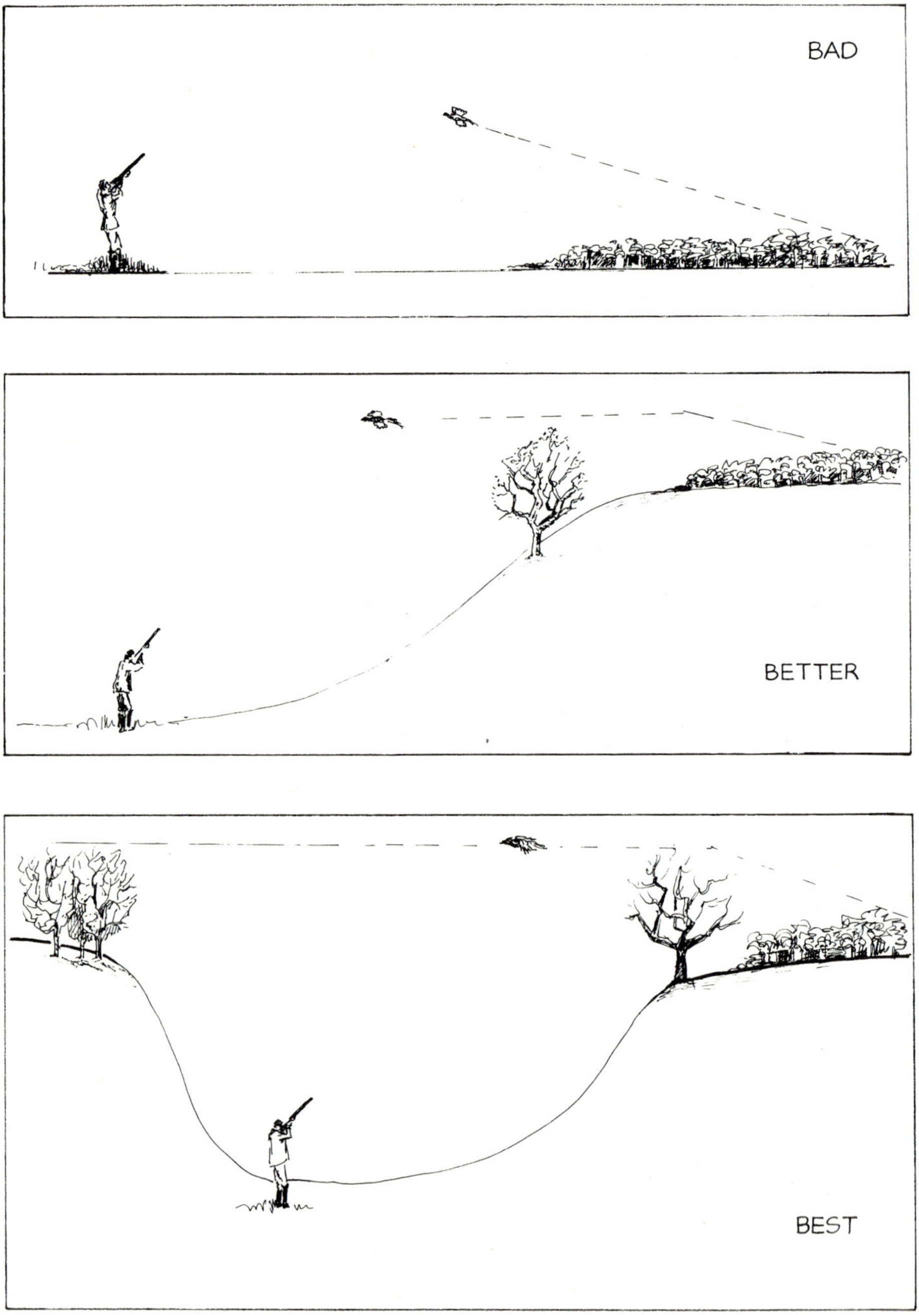

PHEASANTS FROM KALE

and some kind of grain. After the harvest the stubble was left alone until the kale had been eaten off and then the whole area was ploughed. This is a splendid way of holding game on the ground but modern farmers seldom approve.

Most shoots buy corn from the home farm, but few farmers plant anything as game food pure and simple and, of those who do, almost none grow it where its position will benefit the shoot. It is a sound plan to plant crops of proven merit as sources of food (buckwheat, maize, sunflowers, artichokes and some of the new 'game mixtures' are examples) in the places where they will serve the interests of the shooting – thus combining the source of food with the holding cover. This must be made profitable to the farmer and a proper business arrangement should be made, but it may not cost the shoot anything. After all, the shoot will spend so much on food for game and if that sum produces more nourishment when devoted to growing buckwheat or the like than when buying sacks of grain, all hands should be happy, and the added ability of the ground to attract game is a bonus.

This may be the place to point out that all farmers are not inspired by the same motives. Some, and they should not be reviled, work at home to make money which they spend elsewhere; others enjoy the living which their land provides and are quite prepared to fall in with schemes which may reduce their paper profits if they yield rich rewards in sport; and these are the shooting landlords to seek out and cherish.

Long-term improvements

Look at the drawing on page 50 which shows 'A good, but not perfect, covert' and please do not exclaim that anything on that scale is right out of financial range and far too much like hard labour. So far from costing money this covert will make a profit, and if forestry is treated as a recreation the woods will improve without toil.

Notice the perimeter hedge on the drawing. It gives cover from view and cuts off draughts at ground level; but it is important to realise that if this hedge were replaced by a barbed wire fence the covert would be effectively narrower, because the outside 20 yards or so would not be attractive to game.

Just inside the hedge are young Christmas trees among which pheasants can nest; some are lifted and sold each year to bring in cash but one or two rows have been allowed to grow to a height of 15 or 20 feet to shelter roosting pheasants from the wind and prying eyes. The main body of the wood consists of oak or beech trees, which are felled every hundred years or so and replanted. Enough light will probably penetrate to allow a

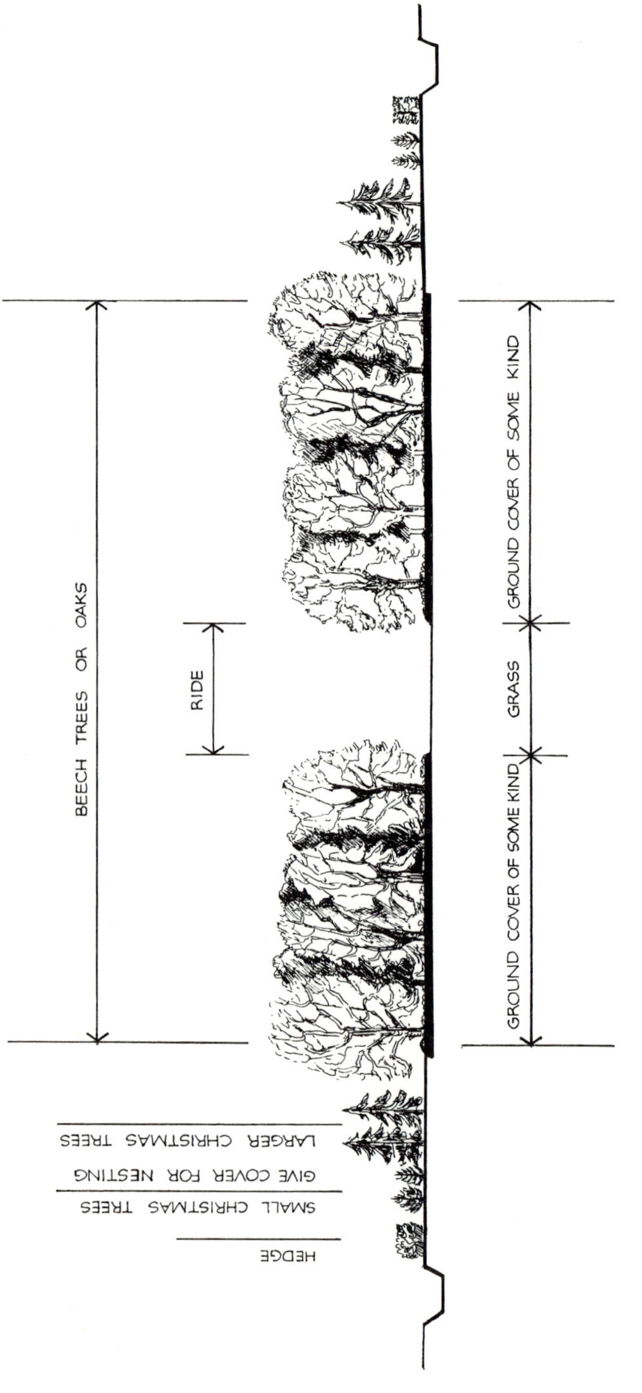

CROSS SECTION OF A GOOD, BUT NOT PERFECT, COVERT

moderate growth of weeds and bushes on the ground and there may be a middle tier of sweet chestnut coppice, but hazel coppice was more usual in the old days. If the wood is of any size there will be a wide central ride and possibly a clearing in the middle.

There are thousands of such woods in existence because they are sound financial investments. No doubt the men who planted them wished to beautify their property, to have a convenient source of assorted timber and to add to the amenities of the estate but, first and foremost, all such woods are shrewd investments. That they make ideal homes for game is due to chance and the amazing adaptability of the pheasant; and that the sporting rights of such woodlands have recently become more valuable than the timber is nothing but a twist of Fate.

A shooting tenant may not be directly interested in the value of woodlands as investments but relations with the landowner will be easier if he understands the fundamentals. It is important to know that the men who planted these woods were very good judges of the quality of soil and seldom put trees where crops would be more profitable. They were also expert at draining land and if they dug a ditch it is long odds that it pays for its keep.

The hardwoods, that is to say the oaks and beeches, are for capital growth, the conifers which were planted to shelter the infant hardwoods brought in cash as they were removed progressively, and the coppice of hazel or sweet chestnut paid a steady dividend. The timber from the odd trees such as ash, larch, willow, hornbeam, crab apple and sycamore was used either on the farm or within the community and the evergreen bushes were valued as household decorations. The wide ride in the middle of the wood was the route by which the timber was extracted and it also gave working space to the men who made hurdles from the hazel coppice or split and worked the chestnut. This was also the original purpose of the central clearing, and the fact that these clearings make ideal places for releasing hand-reared pheasants is no more than blind luck.

This utilitarian investment offers a lot to game, particularly if a few minor modifications are made. The acorns and beechmast are major items in the diet of pigeon, mallard and pheasants; the whole covert provides shelter from the weather and from view; and there are potential sites for nesting along the perimeter hedge and in the wide ride and central clearing, both of which allow game to relax and preen in peace. As long as the ground cover is not too dense, it hides birds from their enemies but allows them to escape on foot or to fly if hard pressed.

If the ground cover is a mixture of all manner of grasses, weeds, bushes, flowers and shrubs it provides food of some kind over many months of the year; and notice that this is not so if there is a uniform growth of, say, brambles. The chief value of isolated evergreens, such as holly and laurel, is that they cut off draughts and give shelter at all times of the year; they also reduce the range of vision, make useful escape cover which may well not be noticed by a predator hunting the perimeter hedge and, on shooting days, they make useful local flushing points.

The value of a stout perimeter hedge which excludes trespassers, breaks the wind at ground level and gives cover from view has already been explained; but it can also be a considerable source of food. On balance, a quickthorn hedge is probably the best, although this can be debated. If it is kept neatly trimmed at chest height there will be no berries, but if part of the hedge were trimmed and the rest allowed to grow, it would yield fruit which is eaten by game. And if the trimmed portion were left untouched another year, when the tall part was cut back, there would be an annual addition to the food supply although the hedge would be equally well maintained at no greater cost.

I do not pretend to know how many bushels of berries are provided by 100 yards of thorn hedge, but the equivalent amount of wheat would cost an appreciable sum. And when one considers how many miles of hedges there are in the aggregate, and how many extra pheasants would come to eat the berries if only the hedges were trimmed with the welfare of the game in mind, one gets some idea of the scale of improvement which can result from a little thought.

The merits of planting shrubs which bear berries both as alternatives to thorn hedges and within woods are discussed in Chapter 7, together with other means of providing cheap food for game. The following true story describes a very effective solution to the problem.

The new owner surveyed part of his farm somewhat sourly. There, as prominent as a vintage Rolls Royce on a rubbish heap, stood the ruins of a once magnificent covert through which a biting east wind blew with little to check its passage. That covert had, indeed, seen better days. The trees rose from a quagmire of bare, churned mud, every bush and blade of grass had been destroyed by rootling pigs, every ditch was full of debris and floodwater stood in pools. This desolation was surrounded by the remnants of a hedge which had not been trimmed for generations.

It was a depressing scene, but two things kept hope alive: most of the trees were well-grown hardwoods, and valuable if they could be saved from the floods, and the covert had evidently been planted by someone well versed in the twin arts of making pheasants comfortable and presenting them to perfection. The trees were a clever mixture of species, each placed to give of its best. Conifers for shelter, holly and rhododendron for warmth and larches for roosting, were scattered among the oaks and beeches which made up the bulk of the wood. The very shape of the covert, an equilateral triangle, showed a touch of genius since pheasants could be driven downwind no matter whence it blew, and the beaters would be brought closer together as they approached the flushing points near each apex.

The flushing points themselves had been destroyed by the pigs, but this was almost an advantage. In accordance with the fashion of the time, they had consisted of patches of snow berries and privet but shrub honeysuckle (*Lonicera nitida*) would serve the purpose better. Traces of the hazel and sweet chestnut coppice which allowed the birds to climb easily before clearing a ring of tall trees on the perimeter could still be seen.

Urgent action put the drainage to rights, but after that things rather ran themselves. A pensioner with an unusual axe arrived, sharpened a pencil with his billhook, did some calculations on the palm of his hand and suggested that he should be allowed to show what he could do with the long-neglected hedge.

A trial length having been selected, well away from the lane lest failure bring ridicule, the pensioner went to work and both his craftsmanship and the result were a joy to see. Twisted thorn trees 20 feet high yielded to his axe and fell just where they were wanted on the line of the hedge without being cut right through. Once the billhook had cut away surplus boughs, another tree was felled to mingle with the first and, rather unaccountably, the two seemed to weave themselves together without much human aid. As the work progressed a dense mass of living wickerwork replaced the row of isolated trees to which the original hedge had degenerated.

When the growth of summer had thickened the tangle and masked the scars of cutting, the hedge looked much like any other, but neither man nor beast could push through or wriggle under; and in its shelter both the wood and many pheasants have prospered. That pensioner was an expert, but hedges can be cut, laid, staked and bound by any determined man who applies himself to the task. A novice's work will be slow and of a lower standard than that of an expert, but it will serve its purpose; half the battle is learning how to put an edge like a razor's on all the tools. The man whose billhook will not sharpen a pencil is making trouble for himself.

The central clearing of this covert had been taken over by seedling birch which was cut down easily enough with a chain saw, and the sale of the timber more than covered the cost. There was no difficulty in establishing grass within the clearing once the birch was out of the way, but inducing any form of ground cover to thrive under the canopy of mature oaks presented problems which were never completely solved. Where gaps in the canopy allowed sufficient light to reach the ground it was found that laurel, mountain ash, the more robust cotoneasters and shrub honeysuckle did tolerably well if planted as well-rooted bushes. If planted as cuttings, they either failed to take root or were overwhelmed by grass.

Where there were no gaps in the canopy, many experiments failed to discover anything more satisfactory than allowing Nature to cover the ground with anything which would grow of its own accord – brambles, wild raspberries, foxgloves and bluebells seem to be the dominant species. Year by year this natural growth covers a greater area, but progress is slow.

Other satisfactory woods

Until now we have dealt with woods which are sound investments in themselves and which produce game as a by-product, but there are others which are almost better from the shooting man's point of view. One such wood consisted of 150 acres of almost impenetrable scrub in such deplorable condition that two stout-hearted amateur keepers obtained the shooting rights for a very small sum. The history was that this 150 acres had been normal woodland until the hardwoods had been felled in the First World War, and the whole place had been allowed to go back to jungle until the amateurs took over in the late 1950s.

They set about the keeper's work in the way which has been described in the previous chapters, but they came upon a major snag. There were pheasants, pigeon, woodcock and snipe in the wood, and a small lake held some duck, but no one could shoot them. All a man ever saw was something which might have been a feather duster passing above a dense canopy of twigs. Clearly something must be done to provide rides from which the sky could be seen and, equally clearly, those rides must be cut in the correct places at the first attempt. It proved to be surprisingly easy.

At the turn of the century this wood had been the scene of some top-notch shooting and research discovered an old Ordnance Survey map which showed the layout of the rides at that time. These were located on the ground and everyone who passed along swung a billhook at obstructions. In a short time a man could walk freely along all the principal rides, although there was not a single decent stand for a gun. Then there was a great studying of maps and of the ground to decide the directions in which the pheasants could be driven and where the guns should be placed; the theories were tested by trial and error and finally an enthusiast spent several days with a felling axe clearing two experimental shooting rides. One was an instant success but the other never quite came off; nevertheless the pattern was set for the future – if you were not satisfied with your field of fire the remedy lay with your own axe or chain saw.

In much the same way persistent, light-hearted slashing with kukries, machetes, sabres and all kinds of cutting implements, ancient and modern, tamed the impenetrable thickets until the wood was something of a show-piece and the bag paid the shooting rent several times over. It was found that this second growth scrub held pheasants better than ordinary woods because the great variety of bushes, trees and weeds offered food of some kind at all times of the year.

Several important lessons were learned during the rehabilitation of this wood, or perhaps forgotten lessons were re-learned; and among them was the folly of galloping willing horses to exhaustion. As long as manual work is fun, untrained amateurs can achieve a great deal, but no non-lumberjack can swing a felling axe for a full working day. If he tries to do so he will come to hate the sight of an axe and he will not go near the woods; but if he stops before he becomes over-tired and does something useful which uses different muscles, perhaps burning the debris, he can work indefinitely.

A particularly dense part of this wood was never thinned because it provided a sanctuary which ensured that some game, at least, would survive to breed despite poaching and over-shooting. It also made a safe haven for predators, and a purist would claim that over-shooting and poaching should be eliminated by better means, but the scheme is not without all merit.

Presenting pheasants well, that is to say putting them over the guns in a sporting manner, from a large area of second growth scrub is difficult, and the cost of beaters is high, but in this instance the results were more than satisfactory. By taking plenty of time, learning from experience and siting each stand with loving care, some good and testing pheasants were seen in this wood. With the original 150 acres as the centre the shoot was expanded to 700 acres and 1000 hand-reared pheasants were released, of which about one third were recovered, each year. As this equals the average for the whole country, those self-taught amateurs did pretty well on land which was so bad that no one else wanted it.

Stands of dense conifers

Large blocks of conifers should be assessed on their merits but it is important to know that conifers produce no food for lowland, English game. They may, and do, give nesting places, peace and quiet, shelter from the weather and protection from predators, but the game must seek food elsewhere – and everything else stems from that.

If the trees are so dense that the ground below is without growth, only the rides can furnish natural food. Hand-reared pheasants have been released successfully on these rides but the birds must rely entirely upon food given by man, plus whatever they can find in the surrounding fields.

Things are more promising if the conifers have been thinned so that enough light penetrates to allow grass and weeds to grow on the floor of the forest. Such places have value as nesting ground particularly if fields or deciduous woods are reasonably close.

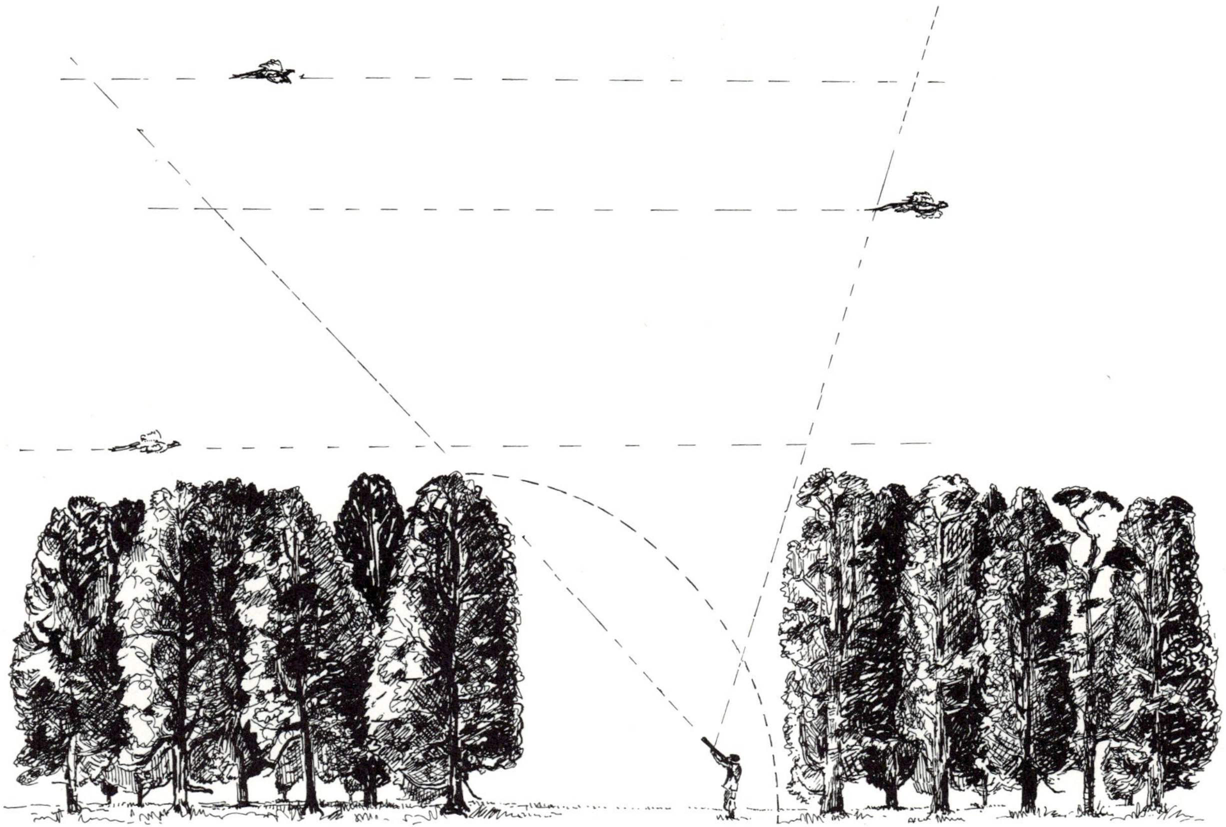

A SHOOTING RIDE WHOSE WIDTH EQUALS THE HEIGHT OF THE TREES.
NOTICE HOW THE HEIGHT OF THE BIRD AFFECTS THE TIME IT IS IN DANGER.

Forestry and presenting pheasants well

To 'present' a pheasant is to show it to a man with a gun, and to present it well implies that the bird is unlikely to be harmed unless the man is a useful shot. Some clearing of rides, planting of flushing points and forestry in general is done with the intention of improving the presentation of the pheasants, and these notes are confined to that aspect alone. In practice, this work is done in conjunction with the improvement of the habitat, which is the reason for its inclusion in this chapter.

Pheasants flushed within a wood climb to clear the trees and then fly more or less level, so if the trees are tall the birds will be higher and harder to hit. If men standing within the wood are to have any chance at all they must have room to swing their guns and they must be able to see the sky. That is to say they must be in a 'shooting ride' – but how wide should that ride be? I am convinced that the correct width of a shooting ride is not any particular distance but is the height of the trees in front. Then the man first sees the bird when it is 45° above the horizontal and must fire when it is either in front of him or directly overhead. As soon as the bird is a few yards past the gun it is hidden by trees and safe. That is as it should be, I believe, because for no logical reason I hate to see men turn round and fire at driven game which has passed above them.

The next question is what is the proper distance between the guns standing in a shooting ride? And the answer is that 50 yards between guns is the absolute maximum, but this can be reduced to as little as 20 yards if the men are sufficiently well disciplined not to fire at their neighbours' birds.

If the shooting rides run east and west, it is always doubtful how far south of Ride A Ride B should be cut. I have never known much good result from having B less than 250 yards or more than 600 yards from A. Trial and error would give the best solution, if it were possible, but if I were compelled to guess I should settle for about 400 yards.

In a big wood the pheasants will not allow themselves to be driven tamely forward. They are very clever at escaping round the flanks of a line of beaters, so there should always be a 'walking ride' on the sides of each drive. This need be no more than a track, but something which gives the walking gun a sporting chance at anything going back or sideways is better.

If the pheasants were allowed to run to the hedge at the very end of the wood, and to rise from there, they would not be compelled to gain any degree of height and might be no more than a few feet above the ground when they reached the guns. So a length of sewelling, which is string with rags tied to it at intervals, is run right across the width of the drive about 50 yards from the end. This sewelling should be supported on sticks about 3 feet above the ground; the wild pheasants will not dare to run under such an unnatural object (although the hand-reared may unless a man in hiding twitches it) but will rise within the wood and will climb to clear the trees. And this is where we help them by providing the straight, easy flight path shown in the illustration.

A pheasant uses a lot of energy fighting its way up in spirals through trees, so it has that much less power when we want it to be at its best. A gentle, straight climb is much less exhausting. Then it will fly level to gain speed and will rise again when it sees the guns; and if it has the strength to increase speed while it is climbing it will tax the skill of very good shots indeed.

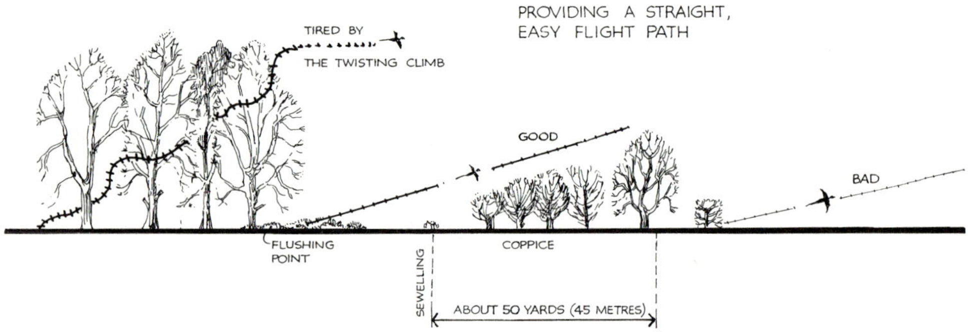

Flushing points

Each drive should have a major flushing point at the place where you want the pheasants to leave the ground. This may be at the brow of a hill, before a certain ride is reached within a wood or, as has been explained, 50 yards or so short of the end of a covert. Essentially, a flushing point is a clump of bushes into which the pheasants will run and hide when alarmed by the beaters and from which they will be flushed later. For perfection, major points should be equilateral triangles with one apex towards the guns and the base parallel to the line of the beaters. A drive of average width might have two, or possibly three, such triangles just short of the sewelling.

It is a mistake to deny the value of flushing points although it is possible to get pheasants into the air without them. The point is that fewer pheasants will steal off sideways or break back if they know that there is a good hiding place in front and, more important, those which rise from the flushing point will be harder to hit. The bushes which make up the flushing point may be snow berry, laurel, small rhododendrons, small Christmas trees or even brambles kept within bounds; but for a new planting in a place where there was plenty of light I should be inclined to choose shrub honeysuckle (*Lonicera nitida*). This should be planted in rows 4 feet apart and with 4 feet between the bushes in each row. There is no need to buy such shrubs from a nursery because, 'Will you let me keep the clippings if I trim your hedge?' normally yields ample cuttings which can easily be rooted in the garden before being planted in the woods.

So much for the major flushing points, but the more numerous the minor points in the course of each drive the more evenly will the pheasants rise. Minor points are any small patches of cover in which birds may squat down and hide; they may be brambles, isolated laurel bushes, holly trees or the like, whose primary purpose is the improvement of the habitat.

On balance it pays to bear in mind the work which is intended to improve the presentation of the pheasants but to keep the options open until experience has shown where the birds congregate and in which directions they tend to fly. The drives should be arranged in a logical sequence, each supporting the others; in general they should be from the boundary of the shoot rather than towards it, and both downhill and downwind, but pheasants cannot be driven in any direction they dislike and the planner must bow to their wishes. There is, however, one firm rule about rides: there should never be a clear view along a ride from the outside of a covert. Always arrange a bend in a ride where it joins the perimeter hedge, thus cutting off cold winds and inquisitive eyes; build a wall with bales of straw as a temporary measure if necessary.

The booklet *Forestry and Pheasants*, published by the Game Conservancy, is a mine of information on this, and related subjects.

7
Cheaper Food for Game

The by-products and wastage from agriculture and forestry have fed pigeon, wildfowl and pheasants for generations. When these did not suffice, man usually gave them wheat or barley because those were the cheapest forms of nourishment available in bulk; but their price has trebled in the last few years and the hunt is up for cheaper foods. Let it be said at once that there is no simple, ready-made solution because those who rear turkeys and chickens have been studying the problem for years and have exhausted the possibilities. A saving of one half of one per cent is a triumph in those circles but is of no importance to the manager of game; if he is to make progress he must either break new ground or introduce new factors. Within limits both can be done but the problem must be stated clearly if the value of possible innovations is to be assessed.

Let us consider the size of the matter first. A pheasant requires 2 ounces of food each day, as does a duck, so a ton of wheat will last 600 of either for a month. Pigeon and partridges eat slightly less but that does not affect the figures. A significant difference is made to any shoot by 600 extra game birds, but one ton of wheat will only feed 200 pheasants through the great dearth of January, February and March; and it will satisfy no more than 100 from 1 October to 1 April. Of course, man only gives what the birds cannot obtain unaided from the natural sources, but my point is that if substantial stocks of birds are to remain on the ground the manager must think in terms of tons of food, not of pounds or even hundredweights. Also, and this is all-important, food must be available throughout the year. That there was ample natural food for 1000 pheasant in November will not make an atom of difference to the number nesting the following April if they could find nothing to eat throughout February.

Now let's think about the chief sources of natural food in winter; these are the stubbles until they are ploughed, acorns, beechmast, berries and the seeds of weeds and grasses. Admittedly, all game eats a host of other things but these are the mainstays; so what can be done to add to their weight? To delay ploughing the stubble is bad farming so that cannot be considered; to plant oaks and beeches would help our grandchildren rather than ourselves and no one wants more weeds, so berries are the best hope.

At this point we run into one of the great obstacles in the path of clear thinking, which is the half-truth that has been repeated so often that it is accepted as true beyond all question. And here the half-truth takes some form akin to, 'Is it possible that you do not know, Sir, that although pheasants roost in woods they obtain their food from the fields?' Most certainly pheasants, duck and pigeon all obtain a great deal of food from the fields, but how many tons of acorns and beechmast do they find in each acre of woodland? And how many tons of berries does each mile of thorn hedge produce? I do not pretend to know the answers but they must be large quantities, and hungry pheasants will gobble them all up.

I have already described how ordinary hawthorn hedges can add materially to the supply of food if they are trimmed in rotation, and that this neither adds to the cost nor lowers the standard of maintenance. There are many woods and shelter belts which are surrounded either by barbed wire alone or by wire with little more than a token hedge inside it, and the value of the covert from the game manager's point of view would be increased if a robust hedge existed. So I suggest that a steady programme of planting hedges around such woods and shelter belts should be undertaken. The work would be chiefly for the benefit of the habitat, but if berry-bearing species were planted, not only would the supply of food be increased but selection could ensure that each kind ripened at a different time of the year. It is, in fact, possible to plant hedges of thorn-bearing shrubs whose berries are ignored by birds until about February and are then eaten with zest by pheasants. I think that that is something of a triumph: the main object, which was the improvement of the habitat, was achieved, but a by-product is additional, free, natural food at the time of greatest dearth. The Latin names of the two shrubs which did this are *Berberis thunbergii* and *Berberis thunbergii atropurpurea*. Both are varieties of berberis which shed their leaves in winter, but it should be remembered that some other kinds of berberis are most unsuitable.

By following the same line of thought it is not difficult to provide cost-free food for game as a by-product of several things which will be done in any case. Sweet chestnut coppice is a facet of forestry which pays a steady dividend, so a thoughtful man plants it between the flushing points and the ring of tall trees on the perimeter of a wood over which the pheasants will be driven on shooting days. Then the coppice pays a dividend, the pheasants are presented better and the nuts themselves are eaten by the pheasants. At least the Game Conservancy and many other people tell me that this is so, although I have yet to find a sweet chestnut in a pheasant's crop, and I have an ingrained suspicion of all such lore until my own eyes have confirmed it.

This distrust dates from the moment when I first saw flying-fish flap their wings and visibly increase their speed by so doing. As a child I had often read that flying-fish gained all their pace in the water and glided on fixed wings in the air, losing speed steadily; so when I saw this was untrue in this case at any rate, I formed a low opinion of pseudo-naturalists who copy each other's mistakes. And my study of the berries eaten by pheasants has done nothing to change it.

There are, indeed, major differences of opinion about the kinds of berries which are eaten, almost every variety being praised by some authority and condemned by another. Also, the whole subject is bedevilled by the fact that ripeness affects the issue; hungry pheasants will ignore certain berries for months and then suddenly feast upon them, presumably because some particular stage of ripeness has been reached. These notes are therefore confined to species which I have seen pheasants eating in large quantities.

Apart from hawthorn berries, which are eaten in the autumn, all the cotoneasters are welcomed and have the advantage that each variety ripens at a different time, some quite late in the winter. Even so, the varieties which either hug the ground or cannot support their own weight should not be planted in the woods because they will be overwhelmed by grass and weeds. The big, robust, quick-growing varieties should be selected, even though they may need cutting back occasionally. A little work with a hedging knife is a small price to pay for the food they provide and their ability to suppress rivals by out-growing them. The frail varieties which bear masses of berries when trained along garden walls are best avoided if they cannot fend for themselves in the wild.

Mountain ash (rowan) provides food which may be out of the pheasants' reach but snow berries have some merit; privet has a reputation for value which I suspect but pheasants eat elderberries and the seeds of broom. Shrub honeysuckle (*Lonicera nitida*) gives no edible fruit but young pheasants flock to it in summer because it harbours so many insects; it also has the advantage that rabbits seldom nibble the young bushes. Ordinary blackcurrant bushes were planted in the woods by the old-timers because the fruit helps to prevent straying, but I believe that a dung heap serves this purpose better.

In warm weather, pheasants are drawn to dung heaps as iron is drawn to a magnet. The insects are the attraction; indeed there is evidence to suggest that these are the favourite food of pheasants in the summer, but young pheasants are also great seekers of sunlight. So the thing to do is to make a dung heap in a sunny place where you want the pheasants to gather; and on a number of grounds it is better to place it just outside a covert rather than in the central clearing.

The case for a determined effort

It is easy to pour cold water on suggestions like the foregoing, for not one single shrub is so perfect that it cannot be faulted. Success is not assured, the work is hard and the harvest is far away; but what is the alternative? To sit with folded hands meekly watching the bills for food mount is to ensure ultimate failure, but a level-headed attack on the problem may bring savings as great as do the acorns and beechmast which we owe to the far-sighted men who planted the trees.

If you intend to plant flushing points, why not choose shrubs which have some feeding value? And the same thought applies to all hedges, windbreaks, random cover within woods or any other planting. The work is not necessarily back-breaking or expensive. After all, most of the shrubs can be grown from cuttings or seeds planted in your own garden; and if only 50 yards of hedge were transplanted each year it amounts to 500 yards after ten years, and all those cotoneasters would produce a lot of food every year without costing a penny. Furthermore they reproduce themselves to some extent.

Much of this planting with a main objective and a profitable by-product is venturing onto untested ground. Some of the brightest prospects may prove to have unsuspected faults, such as forming impenetrable thickets or playing host to agricultural pests, but enough is known of the subject to ensure at least partial success. The sensible course might well be to press ahead with a flexible mind – to experiment, to be prepared to abandon failures and to reinforce success, but not to cross the boundary between enterprise and recklessness. My own experiments taught me that it is unwise to mix varieties too freely, because the quick-growing shade out the slow. Better results come from planting fair-sized patches of one variety of one species. And it is fatal to plant so much that the weeding interferes with more important tasks.

8
Attracting Wild Duck

This chapter concerns inland wildfowl, which is a branch of the keeper's craft that is crying out for research and development. With the possible exception of pigeon, no bird yields more shooting for the work put in than do duck but, and this is the reason why they are not preferred to pheasants, they are so self-reliant and so capable that they may leave the shoot overnight.

A large number of mallard nest in this country and they are always flying about in search of greener pastures. The migrants add to this number in the autumn so that few patches of fresh water are entirely without duck, but it is less widely known that there are relatively few fields which are not visited by mallard at least once in every year. They haunt the stubbles, dibble along most streams, feast in potato fields after the crop has been lifted and investigate any area which is flooded even temporarily; but they come at night and their visits are often unsuspected by men. Persuading them to come to your shoot is chiefly a matter of showing them a glint of water and ensuring that any duck which settle on it find a meal.

The ordinary pattern of a mallard's life is to rest on, or near, water during the day, to flight to a feeding place at dusk and to return to a resting area at dawn. It follows that the guns should be stationed near the feeding place at sunset, while they should try to intercept duck flying to the resting place at dawn. But this pattern is often broken. If the feeding place happens to be water and the duck are not disturbed they may spend the day there and go on feeding at sunset, and then there will be neither an evening nor a morning flight.

Duck are friendly creatures which, like pigeon, prefer to eat in company and are willing to share a meal. When those flying see others eating they may well come down to join them and, if they like what they find, the news spreads quickly, provided that the place is reasonably close to an established flight line. The key to increasing the number coming to any particular place is to put down food regularly in smallish quantities and to shoot seldom; once a month is about right.

Duck will investigate almost any patch of fresh water but their preference is for undisturbed feeding, shelter from the weather, a bank which allows them to come ashore in comfort and a reasonably clear space around them when they have done so. Bushes right up to the water's edge are too likely to conceal foxes and afford no space for preening and resting; an island with shelving banks in the pool is always an advantage.

For the diving varieties of duck the depth of the water is immaterial but both teal and mallard feed either on dry land or in water no more than a foot deep, and such places soon become overgrown by sedge and rushes. The best scheme is to arrange for a landing zone where the water is too deep for reeds to grow; this calls for a depth of at least 3 feet, but 4 feet is safer; perhaps it is the reflection of light from the surface of this landing zone which attracts the duck in the first instance.

If the pond has reeds and sedge with a few bushes on the perimeter to give cover from view and break the wind, so much the better; and if the duck can settle on the water and walk to a barley stubble things are perfect. The best of flight ponds, however, will be trampled into a horrid mess if it is not fenced to exclude cattle, and it is usually possible to allow a corner of the pond to protrude under a fence so that the cows can drink.

There may be no such pool ready-made on your shoot but the countryside is dotted with tens of thousands of disused ponds, abandoned gravel and clay pits, ornamental lakes and the like which can be made attractive to duck by working in the directions already suggested.

Making a flighting and breeding area

As long as there is a place which is even remotely suitable, it is not difficult to contrive a pond to which duck will flight in the evening; but the same place often gives the opportunity of providing a nesting ground, and the work is so similar that one description will serve for both.

The first essential is a patch of open water at least 3 feet deep with shelving banks; then there should be a dibbling and feeding area with water no more than a foot deep. Shelter from weather and view is essential for nesting but is not vital for feeding. Ducklings, however, must have areas where they can rest, relax and fossick about in search of food. A mudbank will serve this purpose but ducklings are so vulnerable to winged predators that something similar which provides cover is far better. A series of bays is better than a straight beach because a duck with a family is inclined to be aggressive and it is as well to keep the broods apart.

No great effort is required to visualise a fairly level piece of ground flooded to a depth of a few inches with tussocks of grass, reeds, iris and scrub willow growing thickly. If that area were sprinkled with fair-sized bomb craters it would do well both as a flight pond and as nesting ground. The size is not important; obviously the bigger the better but the area of a tennis court is not to be despised.

The first reaction is usually that such a place could easily be made by damming a stream; a vision of a telegraph pole laid across the waterway supporting the tops of sheets of corrugated iron which have been driven into the earth leaps to the mind. Imagination insists that that would make a fine pool, and so it might for a day or two. Unhappily, very few dams built by amateurs last for twelve months. We will come to the reasons later, but the golden rule is to dig rather than to dam.

The result is the same whether the digging is done by one man with a spade or by sophisticated earth-moving machinery, and the latter may be available either free or for much less than the advertised cost of hiring. Operators must be trained, machines must be tested and firms sometimes demonstrate plant to potential customers, so a man who puts the word around may hear something to his advantage.

Most cheap excavation, however, is done by a contractor's machine which would otherwise be standing idle. Perhaps a drag-line has been working near the shoot on some major project and has finished its immediate task but will be needed again in a few days' time. Rather than see it stand idle, the contractor may allow it to work on the shoot at a reduced price.

In such a case the best plan is for the contractor's representative to visit the site with the landowner and the shooting tenant to agree exactly what work should be done. Stakes which can be seen from the operator's cabin must be put in to show the line and level of the excavation and where the spoil should be dumped. A big machine can only make a profit when it is working in conditions which give it a clear run, so the tenant may have to abandon some refinements of design to allow the digger to go at the job bald-headed.

The tenant should be present when the machine arrives to ensure that it reaches the site without getting bogged down and can work without interruption. He will probably have to replace the marking stakes many times, and the number of idle spectators, all apparently bent on suicide, who manifest themselves from an empty landscape is downright amazing.

Some shooting men will remember the ease with which the wartime Sappers blew enormous craters with explosives and how their help facilitated the digging of gunpits and the like. Given the necessary skill and equipment, explosives may well provide the cheapest, quickest and easiest means of making pools in a swamp, but they do not tolerate mishandling. It is better to forget wartime skills and leave the blasting to those who have all the proper permits.

Dams

Home-made dams usually fail, not because the dam itself breaks but because water seeps underneath or round the ends. Alternatively, because the overflow is too near the dam, eddies form and erosion does the rest. Making a dam of any size which is expected to last really calls for the advice of an experienced Civil Engineer, but there are exceptions. Men brought up in the fens of Lincolnshire or in the paddi fields of the Far East have something in their bones which enables them to build earth dams which endure.

Even with ideal, tenacious clay they are very careful. The width of the top of the dam is at least the depth of the water plus a foot. The sides are either sloped at one foot vertical to three horizontal, and grassed to resist erosion or consist of sheet piling driven into an impermeable stratum and stoutly braced. The overflow always looks over-elaborate to a layman, since experts ensure that it is amply big enough and erodes nothing near the dam. In dry weather the finished work gives the impression that it was fathered by a nervous incompetent – obviously twice as strong as it need be and grossly extravagant; but it withstands the floods which destroy the cut-price jobs.

Every pool made by damming a stream should have a sluice gate (sometimes called a penstock, hatch or sluice valve) by which the pool can be emptied quickly. Its purpose is to flush away the silt deposited in the pool which would otherwise fill it completely in time.

Feeding a flight pond

An apparently insignificant pool which is correctly fed will yield better shooting than an impressive lay-out kept on short rations. The snag is that the most effective feeding routine consumes a great deal of time and can become very expensive. The problem is always to find a sensible balance between perfection and the best which is possible with the men and money available.

Let's consider the ideal first and then examine economies. We hope that the duck will be on their resting ground during the day. As night approaches they will fly to their feeding places, and let us suppose that a few have our pool in mind for tonight. We want those duck to find an ample meal but to return to the resting place at dawn and to come back to our pool, with their friends, tomorrow.

As the number of duck increases more food must be put down, but it is a mistake to give too much. Not only will it be stolen by rats and moorhens if it is left lying about but the duck, confident of obtaining a meal, may come loafing in at midnight instead of hurrying to arrive at the proper time. In extreme cases they will not leave the pool but will spend the day there; and then there is neither a morning nor an evening flight and the number of visitors does not increase.

For perfection the keeper would put down the right amount just before the duck arrived. This is about 2 ounces per bird, and if an appreciable quantity were left in the morning it would indicate either that the duck had been disturbed in the night or that he had been over-generous. But daily feeding by hand is impossible for week-end keepers, and we will return to that problem.

When deciding what food to give, bear in mind that barley and pig-potatoes are the traditional baits for duck, and mallard are fond of acorns. 'Tail corn', which is the low-grade grain from threshing, is a great stand-by and unthreshed bales or sheaves are valuable. Corn merchants always have a proportion of spoiled grain and it is said that brewers sell used grain which duck enjoy.

The pig-swill from hospitals and army camps is usually removed by a contractor from whom it is sometimes possible to buy it in the quantity required. This serves its purpose both with and without an addition of grain.

Apart from these sources there is a whole range of 'condemned' food and fruit. Millers, bakers and the manufacturers of biscuits and breakfast cereals sometimes have 'condemned' foodstuff; as do all those who deal in fruit and vegetables at any point between the docks and the local greengrocer. Bananas, currants and apples are especially popular with duck.

The need for daily visits to the pool would be removed if one of the Parsons Automatic Feeders, described in Chapter 3, were installed. This is the only machine of its kind on the market and it is a void which is crying out to inventors. It should not be difficult to devise some home-made gadget which upset a bucket of corn automatically every twenty-four hours, but no one has done it yet.

Hand-reared mallard learn to feed from ordinary hoppers but it is doubtful if those bred in the wild do so.

Feeding a nesting area

As with other species, it is a great mistake to stop feeding as soon as the shooting season ends. If a flight pond is fed lightly, the duck will remember and return the next winter with their friends. But the big profits come from feeding a place which is both a flight pond and a nesting area. Two or three broods do not require much in addition to the natural food and they are wonderful allies.

They dress the place, attract migrants and passing duck and speed up the process of making the flight pond popular. Many a successful pond was started by hatching mallard's eggs under a broody hen, establishing the family in the potential nesting area and going on from there.

Wildfowl and predators

In the ordinary way, mature duck are not particularly vulnerable to predators and the principal danger is that they may be stalked by foxes when either feeding or resting on land. As we have seen, the ideal flight pond or resting area has an island where duck are quite safe unless ice forms on the water.

Nesting duck are, however, easy prey for four-footed predators and those in eclipse are defenceless. The methods of controlling these predators have already been described and they are very rewarding when used to defend nesting or moulting duck.

Unhappily, both eggs and ducklings seem to be the favourite food of many winged predators and losses are heavy if man does not intervene. Carrion crows, the greater black-backed gull, the herring gull and to some extent the lesser black-backed gull are major enemies. They all search for nests, eat eggs and snap up ducklings. Whether rooks and jackdaws are as innocent as some would have us believe is doubtful, but jays and magpies certainly kill ducklings and moorhens probably eat eggs.

Two lines of defence are possible. All the birds mentioned may be shot by 'authorised persons', but the gulls come in such numbers that losses will be heavy unless both nests and youngsters are well hidden. Cover and camouflage can do a great deal; all natural growth like reeds, sedge and low bushes should be encouraged, nesting baskets can be purchased and nesting boxes can be made from scrap timber or empty drums suitably disguised.

That pike and rats are the bane of ducklings has already been mentioned, and one mink will wipe out several broods, but herons, which are protected birds, also eat ducklings.

I was told as a child that foxes had almost hypnotic powers over mallard and that the medieval wildfowlers knew this and trained small red dogs to lure duck into decoys. I confess that I classed the information with the charming of warts and rituals reputed to make your enemy's milk turn sour, but I was wrong about the foxes. At any rate some foxes can bemuse some mallard, and I have seen it done twice. Even so, I did not mention it to a soul until a little red Shetland Sheepdog learned the trick and did it for all to see.

This is what I saw. The duck were on the edge of the water and the fox was some 30 yards away, partly concealed by vegetation. The fox tumbled about as puppies do, rolling over and attracting attention but making no effort to approach. The duck were wary at first but eventually began to edge towards the fox with their necks stretched out and their heads near the ground; I had the impression that they were mobbing the fox in some pre-ordained, ritualistic manner. The spell was only broken when the fox made a short rush and caught one. The Shetland Sheepdog followed much the same plan with semi-tame duck on an ornamental lake.

Morning flight

At dawn, the shooting man is trying to intercept duck flying from their feeding places to wherever they intend to spend the day; and they usually choose major stretches of water. Estuaries, lakes, reservoirs, old gravel pits and the open sea are constantly used. They value peace and quiet when resting and preening but will tolerate a surprising amount of human activity, provided that it is in plain sight with a reassuring stretch of open water in between. For instance, resting duck take little notice of passing ships or of motor traffic near reservoirs. They come ashore to preen, and sleep on land if the waves are big, but they will not frequent places where they are disturbed after going ashore.

For this reason an isolated sand bar or bank of mud is always an advantage, and if it takes the form of an island it is even better. Duck resting on the water are at the mercy of the wind and current; they dislike being blown ashore at the foot of steep places where they cannot land and are none too fond of being carried near to danger and then being compelled to fly back to a safe area.

With these factors in mind it is sometimes possible to make a resting lake more attractive to duck; and when it succeeds the number in the whole neighbourhood may increase greatly and permanently. A notable example of this can be seen in Sussex, near East Grinstead, where a large reservoir was constructed a few years ago. Some duck adopted it, were fed by the kind-hearted public and became semi-tame; but the vast majority are truly wild birds which only rest there by day and add greatly to the number seen at all the flight ponds within a radius of ten miles or more.

It is wise to shoot an evening flight pond seldom (once a month is about right) but the morning flight to an inland lake, such as that just described, should be treated even more gently. Twice in a season would be reasonable, with three times the maximum, and it is obvious that the further the guns are from the water and the stronger the wind, the less will the lake be disturbed.

Conclusions

It might be said that as a flight pond should be shot so seldom, and a resting lake even less often, the rewards do not justify the effort; but that is not the whole picture. Admittedly, the set-piece shoots are limited in number but the work done on the pools adds to the population of the whole district, both home-grown and migrant, so that duck are more often encountered when rough shooting or waiting for pigeon. Also, the conditions which favour duck are attractive to snipe and pheasants, for it should never be forgotten that the pheasant's true home is the fresh water swamp.

A more valid objection is that the money spent on food for the duck might be more productive if devoted to the pheasants, and this is sometimes true. It is a matter for judgment but, in general, pheasants are only chosen because they are more likely to be present when they are wanted. Over the years, duck usually yield more for the cash and effort expended but they are given to being absent without leave. The normal result is a modest addition to the bag with an occasional bonanza when all the other pools are frozen solid but yours is free from ice.

The whole business of inducing duck bred in the wild to frequent your shoot should not be confused with rearing large numbers of mallard by hand, as pheasants are reared, and relying on them for the shooting. Although that is far more expensive it makes an interesting variation because the same pens, brooder houses and equipment generally which is used for pheasants can be used for duck, so if one fails the capital is not wasted. Broadly speaking, mallard are easier to rear than pheasants. They always promise great things but no one has mastered the art of keeping athletic mallard on the property. All too often they either become so tame and fat that they will not fly or they fall in love with far horizons and disappear.

9

The Wood Pigeon

This big, handsome pigeon with a white ring round its neck and conspicuous white patches on its wings must be known to everyone in this country and is the mainstay of most lowland, rough shooting. It is probably the target of more cartridges than are fired at all other birds put together, yet it maintains its numbers without the protection of a close season and with very little help from gamekeepers. No one rears wood pigeon and almost no one puts down food for them, although they help themselves to food intended for duck and pheasants, and the only substantial benefits they derive from gamekeepers are the safe roosting and nesting places provided by the coverts.

I think that the number of our home-bred birds is augmented by migrants from Europe in the autumn, although this is disputed by some, but England has vast, built-in pigeon sanctuaries in the suburbs of her towns which must go far towards keeping the population steady no matter how many are shot. And they are harried relentlessly because they pillage the farmers' crops. Recent studies indicate, however, that guns play only a minor part in limiting the number of pigeon; as with other game, the dominant factor is natural wastage.

Pigeon breed from about April right into August or even September. The nest is a frail platform of twigs seldom more than 25 feet above the ground, two white eggs are laid in each brood, the incubation period is about seventeen days and the male bird helps both with the incubation and with feeding the young. These youngsters, which are called squabs, remain in or near the nest until they are about four weeks old, which may account for the high proportion which survive. They can fly reasonably well at the age of six weeks and soon learn to feed on the ground with their parents.

As juveniles they have a brownish tinge and lack the characteristic white ring round the neck but their build and flight are so similar to that of mature birds that they should never be mistaken for any of the doves or for domestic pigeon. The recognition features should be studied because marked ill-feeling results from shooting racing pigeons or family pets.

Wood pigeon have strong wings and great powers of endurance; furthermore they do not seem to develop any affection for one particular area as partridges do. If they like what they find on your shoot they will stay, if not they may all leave together. I think that they go wherever the eating is good and that they prefer to sleep close to the feeding ground. When travelling about their domestic area they may take any course but they tend to follow certain routes, which can be learned from observation but which may also be deduced by a man who asks himself, 'With the wind as it is what course and what height would be chosen by a very lazy pigeon?' They tend to select routes which are sheltered from head winds and they avoid any climbing which a little extra distance eliminates; thus, they usually fly round isolated hills rather than over them and through a gap in a ridge. On short journeys they find their way by going from one landmark to another, and men can see those landmarks just as well as the pigeon can. This is important because the birds must pass close to the landmarks when the visibility is poor.

The general pattern of a pigeon's day is much as follows: leave the roosting place at dawn and fly to find breakfast, eat and go to a nearby tree to preen and rest. If the food is digested quickly, as is the case with tender green growth, the crop may be refilled several times, with an interval for preening between each meal; and it is this constant movement round the feeding ground which enables the shooter to identify it. Then a closer look will reveal the favoured portion of the field and the routes by which the birds are approaching.

When the days are long, there will almost certainly be an interval with little movement in the middle of the day, but feeding will start again at such a time that every crop will be filled to bursting when the birds fly to roost shortly before sunset. When the days are short, the morning feed may run into the evening meal and then there will be movement around the feeding ground throughout the day.

The favoured roosting places can only be found by observation. All that can be said is that pigeon almost always roost in woods, towards the edge rather than in the centre, and that the preferred side changes with the weather. This last point is one of the key factors when selecting a place for shooting birds coming to roost. The man should bear in mind that yesterday's dormitory may be right out of fashion today, and the hitherto neglected side of the same wood may now be full of pigeon.

Reconnaissance is the only certain way of locating feeding grounds but there are some general guides. The shortened version which I keep in my head is that acorns and beechmast are staple foods and grain is a favourite, so stubble and laid corn are always attractive. Any young, green growth such as clover, cabbage, peas, beans, young kale or the like, is eaten, as are the buds of beech trees; and the nearest thing to a certainty with pigeon is that they will flock to kale if that is the only green crop showing above a covering of snow.

A table showing the main diet and the second choices in every month of the year is set out in Archie Coats's well-known book *Pigeon Shooting* (André Deutsch). This is probably the standard work on the subject and it should be studied by everyone who wishes to shoot pigeon seriously.

Organising pigeon shooting

Let it be said at once that there are a dozen routes to frozen boredom while ostensibly shooting pigeon but that there is only one way of arranging shooting for six or eight guns at a stated place a month ahead which is not likely to prove disastrous. With all its virtues, pigeon shooting is not seen at its best when practised by large groups or when tied to timetables. A single individual who is prepared to seize passing opportunities can have a great deal of worthwhile sport, and this can be shared by one or two companions who will turn out at the ring of a telephone, but anyone who tries to arrange pigeon shooting for a syndicate on the lines of pheasant shooting has an impossible task unless widespread roost shooting is the order of the day – and this occurs five times each year, and no more.

For all forms of pigeon shooting I use an ordinary game gun with normal, standard velocity cartridges and about No. 6 shot. Some men whose opinion I respect use more powerful guns, throwing larger shot from heavily-choked barrels, perhaps a magnum chambered for 3-inch cartridges, but the so-called 'pigeon guns' should be avoided. These are highly specialised weapons evolved for shooting live pigeon released from traps. This is a competitive form of shooting which was once popular in this country and is still practised in Europe. The guns are well suited to their own role but are a handicap in the field.

Widespread roost shooting is the one way in which satisfactory pigeon shooting can be arranged far in advance for all the members of a large syndicate. By common consent, all the Saturdays in February and the first in March are devoted to it. On those days, every shooting man available is detailed to defend some wood over as large an area as the grapevine can reach, which may be more than half the county.

The idea is that the pigeon will be feeding in the fields during the day and will fly to some wood as dusk approaches, but they will be driven off by gunfire wherever they try to settle. In theory, all the pigeon in the district will be flying from wood to wood in search of sleeping quarters but will not be allowed to settle anywhere before darkness makes shooting impossible.

In practice, there is usually a lull in the firing followed by a single shot which disturbs all the pigeon in the immediate neighbourhood. That flock flies to some wood, is firmly moved on, together with others which have found a temporary respite near that gun, and a crackle of firing breaks out from successive coverts. This eventually dies away, only for the sequence to be repeated again and again.

In general, shooters should be inside the woods they are defending and reasonably well concealed. Nothing approaching the perfect camouflage which is essential when shooting over decoys is required, but the less the pigeon can see the better. As the birds circle, and they almost always land into the wind, the man sees them through a canopy of boughs and he must decide whether to fire when they cross a convenient gap or to swing normally, ignore the branches and hope that no timber will be in the way at the vital instant.

I prefer to establish myself, perfectly concealed in a hide, near one of the landmarks on a recognised flight path, rather than inside a wood. The pigeon are then travelling fast, not circling to land, and the man is not hampered by trees as he takes on birds which are not at a disadvantage. A good deal of local knowledge is required to ensure that the hide is in precisely the right spot, 30 yards either way may make the difference between some memorable shooting and relative failure, but even then the pigeon may be too high. To guard against that I arrange a place in a wood to which I can move.

Ordinary roost shooting

If pigeon take to roosting in any wood, they can be ambushed by a waiting gun at any time of the year. The method is the same but, of course, the man does not have the advantage that all the other potential roosting places in the neighbourhood are being disturbed at the same time.

The bag is usually small but some shooting is almost always possible; it is largely a matter of knowing where the pigeon are roosting by maintaining a constant watch and seizing the opportunities. I dislike killing pigeon when they may have squabs to feed; I admit that it may be necessary if they are doing serious damage to some crop but, to me, shooting then becomes an unpleasant duty which has nothing in common with sport.

Shooting over decoys

The big bags of pigeon are always made by using decoys to tempt hungry birds within range of a hide. If Fortune smiles, the shooting may be as hectic as that in the best of stands against driven game, and it is as difficult as any. When the gun becomes uncomfortably hot the wisdom, even the sanity, of persisting with expensive, tiresome, hand-reared pheasants will be doubted; but when no pigeon arrive, no field-sport can be a greater fiasco.

Field-craft, concealment and the cunning placing of decoys are essential, but the primary requirement is the presence of pigeon. If reconnaissance has discovered a field in which many are feeding, an excellent day can be expected by shooters who will go out there and then. When there is no such gathering, good sport is still possible if a fair number are moving about, for decoys may lure them down. But decoys are not magical things which conjure pigeon from an empty sky and when the birds have left the district, wise men do not attempt to shoot them.

The foundations of decoying are that pigeon are gregarious (they always prefer to feed in the company of others) and that they are incorrigibly greedy. Even those flying to roost after an ample meal will come to others apparently eating. They may not land at once by those on the ground, and often circle round decoys before settling in a tree close by, but if a flock changes course to approach decoys and then makes off at speed, the man is almost certainly at fault.

Dead pigeon are the best deceivers, some with their heads propped up on sticks. The rubber imitations of complete birds are good and the moulded silhouettes which fit inside each other are the easiest to carry. A single dummy will not be noticed – about fifteen is the useful minimum – but success brings rewards at compound interest because, as more birds are shot and added to the pattern of decoys, the greater is the likelihood of others being attracted. The best scheme is to leave birds where they fall until a lull occurs and then to set them up as additional decoys. Once forty or more decoys are arranged in a lifelike pattern, it is astonishing to see how many pigeon which are almost out of sight will change course and come down to investigate.

When arranging a pattern of decoys it should never be forgotten that the object is to show flying pigeon a sight which they associate with good eating; so a study of a feeding flock teaches the ideal composition. It will be noticed that most, but not all, are more or less facing the wind, that they are fairly well scattered and that only a few have their heads up. Above all, there is no geometrical pattern. If decoys are ranged in ranks, lying on their backs or reflecting light from the surface of wet rubber, passing flocks will not be tempted to join the party.

A camouflaged hole in the ground is the best hide, but a pill-box made of bales of straw is excellent and a screen of boughs or of camouflaged netting will serve. The essentials are that the man should be hidden from all directions, that he should keep watch by looking through the cover, not over it, and that the hide should blend into the background. A common fault is that the man is either insufficiently concealed or is much too cramped. It is certainly possible to shoot well while sitting down, but it is much more difficult; and if the man stands up to fire the pigeon will probably see the movement and take evasive action. In a perfect hide, the man would be completely concealed and motionless as pigeon approached. He would watch the selected bird over the muzzles of his gun without moving until he slid the butt onto his shoulder and fired before the bird could react.

Shooting over decoys has a vast lore of its own. Like keeping wicket at cricket it is a game within a game and even the placing of the hide in relation to the decoys is a minor art. The basic plan is to put the decoys 25 yards downwind of the hide; then the pigeon coming in to land will offer a straightforward shot well within range. If the wind is blowing across the shooter's front, the shooting will be more varied and the birds will probably be travelling faster; while if the man is downwind of the decoys and has his back to them he may enjoy some testing shooting at oncoming birds which are almost overhead.

Those who wish to pursue the subject could not do better than study Archie Coats's book *Pigeon Shooting*. Just how rewarding it may be, can be judged from the fact that Major Coats once shot no less than 550 pigeon in a single day, and even that record has since been broken. Just think of it, all that shooting for the cost of the cartridges alone.

With other game I should not approve of slaughter on such a scale, since the larder does not require it and a sportsman should be satisfied with less; but pigeon in great numbers cannot be tolerated. They eat such enormous amounts of the crops in the fields that it is essential to keep their numbers within bounds. This is not idle destruction, for the meat is good food, but even if pigeon were unfit for human consumption it would be necessary to kill them to protect our own supplies of food; and of all available methods the gun is the most merciful.

Conclusions

It has been suggested that the game shooters of the past tended to ignore pigeon because plenty of more exciting alternatives were available. This may be true, partly because, in some indefinable way, pigeon lack 'star quality'. They are difficult to hit and good to eat but two mallard dead in the air at the same time stir the emotions more deeply and live longer in the memory. This is even more true of pheasants and partridges, and some say that nothing in the world of shooting is so wildly exciting as driven grouse. Also, a pigeon shooter is not really master of his fate; no matter how great his field-craft or how expert his marksmanship, in the last analysis he must wait passively, like a spider in its web, and if the pigeon do not come there is nothing he can do about it.

That may be why there are few among those who habitually shoot driven game who have learned how to come to terms with the sharp-eyed, unpredictable and frustrating pigeon. That art is more widely understood by gamekeepers, farmworkers and others who seldom shoot driven game; but times have changed and the man who does not shoot pigeon is missing a lot of good sport.

The opportunities of shooting lowland game have been so curtailed, the effective length of the shooting season has been so greatly reduced by the decline of the partridge, that values have changed. In practice, the pheasants provide shooting for no more than two-and-a-half months each year, inland wildfowl add something, but that great stand-by, the rabbit, was almost wiped out by myxomatosis over twenty years ago, although it now seems to be returning. Snipe, woodcock and hares give variety but the only species which exists in sufficient numbers to add significantly to our shooting is the wood pigeon. Even if the nesting season is avoided, although there is no legal compulsion to do so, the season then stretches from, say, mid-September to mid-March, and that brightens life considerably.

It is a mistake to under-rate pigeon shooting although it may mean going back to school and mastering a new craft. I am not alone in believing that the average pigeon is more difficult to hit than the average pheasant or duck; although I freely admit that really high specimens of either outclass the pigeon. I have also noticed that men who can hit pigeon perform pretty well against most other kinds of game; perhaps wrongly I attributed this to the fact that pigeon demand every shot in the repertoire in an unpredictable order.

If they could be relied upon to be present when they are wanted, few shooting men would be seriously dissatisfied with pigeon; but only wildfowl are less predictable and it must be accepted that it is impossible to base the shooting of a large group of men on pigeon. There are frequent, but irregular, opportunities for individuals who can hurry out when Fortune smiles, and sometimes the rewards are golden, but the manager of a syndicate who must either invite all the seven members or none has a thankless task.

10
Rabbits

There are two, very different, opinions of the rabbit, and both are perfectly tenable. One school of thought maintains that it is a pest, an unmitigated nuisance which lays waste the crops, undoes the forester's work by nibbling everything and is a thoroughly bad neighbour which ought to be exterminated. The other runs as follows: 'All that may well be true, but let there be moderation in all things; the rabbit has a part to play in the balance of Nature just as everything else has. Let us avoid high feelings and keep the numbers within bounds but let there be no thought of wiping them out.'

When asked what good was ever done by rabbits, the usual reply is that the meat is a good food, that the fur has value, that they keep the undergrowth in check, exercise the dogs and provide a healthier recreation for the children than does the local dance hall. The more resourceful add that it is better that the foxes should eat rabbits than pheasants and that, no matter what the prices of farm produce may be, no farmer need starve while there are rabbits on his land.

What this boils down to is that a shooting tenant should vote as his landlord wishes where rabbits are concerned; but a man who farms his own land can obtain a lot of good food and sport from the humble bunny.

Let us first consider how to implement a policy of extermination. The short answer is that the best way to rid the land of rabbits is to create a disturbance on the surface so that they run down their holes where they can be gassed in the manner described in Chapter 2. Any mild uproar, such as driving about in a Land Rover, encouraging dogs to range over the fields or discharging fireworks, will serve the purpose as long as you see where the rabbits go to ground. But if a man with a gun walks on either side of a hedge while a spaniel works through the thick stuff, the task may be more enjoyable.

Repeated gassing will kill all but a tiny fraction of the rabbits which live in burrows, but some countrymen have always maintained that there are two distinct strains of one species of rabbit, and that one strain seldom goes underground. I subscribe to this belief and suspect that the surface dwellers are smaller, greyer in colour, more nocturnal in their habits and that they are less vulnerable to myxomatosis because they live in healthier surroundings.

Be that as it may, surface dwelling rabbits cannot be gassed but they can be controlled humanely by the gun or the rifle, by the Fenn Rabbit Trap, the Imbra Trap or the Juby Trap. All these traps kill rabbits outright. Once you have found a live rabbit held in a gin trap by two forelegs, each with compound fractures, you should be able to resist the temptation to use gins. They have been illegal since 1958 and were never particularly effective against rabbits; tales to the contrary are based on the belief that the old is always better than the new.

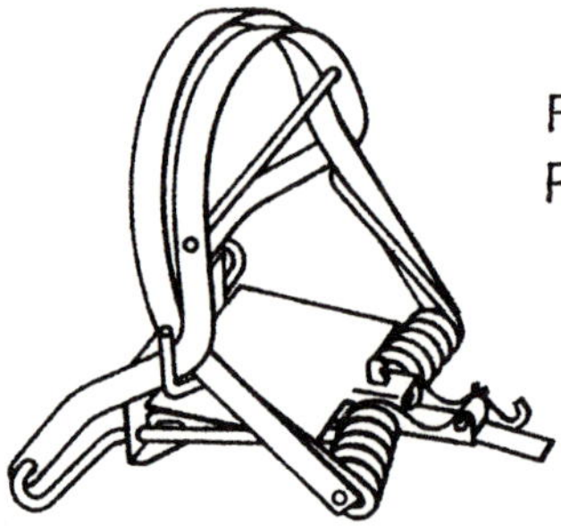

Rabbits for their sporting value

I admit to bias on this subject and maintain that once a novice has learned the elements of using a gun in a shooting school, the rabbit is the best instructor in the field. The need for stealth and quick shooting is evident, estimating ranges and the safe handling of guns become second nature, field-craft is learned subconsciously and the novice who cuts his teeth on rabbits has an enormous advantage over the man who goes straight from clays to winged game. He also has a lot more fun.

Stalking rabbits on a summer evening is not to be despised, sniping them with a ·22 rifle calls for real skill, but a hollow pointed bullet should always be used, and you will remember the day you first shot a running rabbit with a rifle for as long as you live. It does not often come off but it can be done, and for that brief moment Lord Walsingham himself would not despise your skill.

In the ordinary way, rabbits only add variety and substance to rough shooting and walking-up, but there were major rabbit shoots in the past. One of the most memorable was at Blenheim, in Oxfordshire, in 1898 when, after great preparations, five guns shot 6953 rabbits in a single day. This is not a tall story; the names of the men who did the shooting, the drives which were taken and even the number of cartridges fired are all recorded.

Of course nothing approaching that concentration of rabbits would be tolerated by the farming interests now, but it is astonishing that human beings could stand up to so much firing. After all, most of us are feeling the effects after two or three hundred rounds: there may have been giants in those days but even giants of iron must have used very lightly loaded cartridges.

The Game Conservancy has published a green booklet entitled *Rabbit Control*; it is number 22.

11
Partridges and 'Various'

The decline of the partridge is at the heart of both the tragedy and the challenge of managing lowland shooting in this country today. I was brought up with good partridge shooting all round me; had I been asked in the mid-1920s to describe the work of a successful partridge keeper I could have done so with confidence, secure in the knowledge that what was done at home as a matter of routine produced good results. It was comparable with any other facet of agriculture; subject to the weather and so on, if certain things were done commensurate results could be expected. But those methods do not work now, and no one quite knows why. Like many other people I have theories but a cold, hard look at the facts is depressing and the research being carried out by the Game Conservancy is the only substantial ground for hope of improvement.

The overall picture at the moment is broadly this. Over large portions of England, partridges have either vanished or have become very rare in spite of vigorous efforts to check their decline. Some good keepers have just managed to maintain a small stock of birds on ground where they once prospered, and only on the best of land with the best of keepers is there anything approaching the number of coveys seen in the old days. Things vary from year to year but the trend has been downwards for over forty-five years. There was something of a revival in 1976. It might prove to be the first up-swing of the pendulum and the dawn of a new era, but having seen many false dawns I reserve judgment.

Those who are preserving partridges for the first time are inclined to dismiss earlier generations as a bunch of fuddy-duddies too stupid to learn, too infirm of purpose to persist and so accustomed to failure that they will not even make a full-blooded effort. Let me assure them that a great many able, determined, well-informed and resourceful men have devoted much time and work to this matter, although the outcome is still deplorable. This has taught them that caution is necessary; their hopes have been raised and dashed many times but that does not mean that they dismiss the latest chance of improvement as nothing but another Will-o'-the-Wisp – though Heaven knows that there have been more than enough of those. All eyes are now focussed on red-legged partridges and upon trials which indicate that it may be possible to rear large numbers of grey partridges in the same equipment as is used for pheasants; and only time will show whether either path leads anywhere.

I do not believe that the eclipse of the partridge has any, known, simple explanation. It may have been caused by a complex combination of unfavourable factors, but I suggest that it is more probable that there is a gap in our knowledge. If the partridge is to be re-established I suspect that some factor, or more probably factors, affecting its well-being which are unknown at present must first be discovered; and it is probable that better methods of rearing them by hand will be necessary.

But when these discoveries have been made, it is reasonable to suppose that the 'new' partridge will require a habitat and living conditions not totally unlike those in which its predecessors flourished. Some description of what was done in the old days may, therefore, be of value in the future; and whether those methods should serve as a model because they succeeded for so long, or be regarded as an awful warning because they ultimately failed, can be debated.

The land I knew was thinly populated and was farmed on the four course system. The normal rotation of the crops was wheat, roots, barley and clover and there was permanent grass for the heavy horses which filled the role now played by tractors. Almost every farmer had a flock of sheep and a few cows, and it is relevant that the fields were bounded either by stout quickthorn hedges or by big ditches which are locally called 'dykes'.

The keepers ensured that the game was seldom disturbed, peace and quiet was maintained as far as agriculture permitted and the predators were harried to the point of extinction. There were just enough foxes to keep the local hunt happy but any other form of vermin was a rarity. There was not a single crow's nest within five miles of the village and only one magpie's nest was found in the parish in twenty years. The partridges were given a little food in the winter but the cost must have been trivial.

That was the solid part of the keepers' work and the rest might be described as icing rather than cake. Eggs from deserted nests were hatched by bantams, exposed nests were protected, on a very small scale there was an exchange of dummy eggs for real with a re-exchange just before hatching, and there was some attempt to mow round, rather than over, nests in the hay. Whether the exchange of eggs with distant neighbours in order to introduce new blood did much good I do not know, but great store was set upon the custom and care was always taken to ensure that an ample stock of birds was left on the ground at the end of the shooting season. Things varied from year to year but, broadly speaking, the partridges maintained their numbers for many decades with no more help than this and they provided good shooting even in poor years.

If I knew why the virtue departed from the old methods I might be able to solve today's problems. Some of the best of modern gamekeepers, who are the equal of any in history, have seen their partridges dwindle and vanish; and that is why I believe that non-experts stand no chance and should not enter the battle. But that does not mean that the partridges should be abandoned to their fate. If the work I have described is done in the order I recommend, they will benefit, although the main objective was to help the duck, pigeon and wild pheasants. It may enable them to survive until the Game Conservancy produces a solution – which it obviously cannot do without our support.

'Various'

This word is used to describe any acceptable oddment encountered when looking for something else. Thus a woodcock or a snipe which was shot on a day of driven pheasants would be recorded in the game book under 'Various'; and it might be correct to log a goose in the same place if it were shot on a day devoted to walking-up snipe.

There are estates where woodcock are preserved, and the same is true of hares and snipe, but on the average shoot all these species are a by-product of good management rather than a serious objective in themselves. Together with geese, they are all more likely to be found where peace and quiet is maintained, where predators are few, food is plentiful and the habitat is congenial. They are often the mainstay of very small shooting days but few managers plan the keeper's work with any of them in mind.

The reason is that, in this country, it is usually impossible or wasteful to help them very much. If a snipe bog exists you are lucky, so do what you can to encourage worms by scattering farmyard manure and keeping the level of the water just right; but to make such a place is almost always bad farming. Much the same line of thought applies to all the other uncommon species.

12
Wild Pheasants

No statistics are available but it is probable that more cartridges are fired at pheasants than at wildfowl, partridges and grouse put together. As a result of the decline of the partridge they have become the mainstay of formal shooting in England yet there are more misconceptions, fallacies and blind prejudice about pheasants and their shooting than are associated with any other branch of the sport. Let me try to separate the grain from the chaff; it will not be easy because I, like many others, am a partridge man at heart who has been compelled to concentrate on pheasants by force of circumstances.

Up to the early 1920s, partridges were the bread and butter of English shooting; they provided the sport week in and week out. On each shoot there were a few days of driving but most of the time was spent walking-up on quite a small scale, often by two men with dogs. Then, although partridges were the main quarry, the bag always had a proportion of rabbits, snipe, duck and 'various'. The chief requirements were a knowledge of the ways of game, skill with dogs and the ability to locate and come within range of game, rather than marksmanship. As the season advanced, progressively greater stealth and field-craft became necessary, and it should be noticed that this minor shooting employed no beaters, caused few casualties and disturbed little ground.

The pheasants were an occasional banquet, as distinct from an ordinary meal, when a lot of cartridges were fired by men who abandoned cunning and shot at birds driven over them by hired retainers. Then the emphasis was upon skill with a gun, and field-craft was at such a discount that only towards the end of the season was it necessary to distinguish between a cock and a hen. But such days were rare; twice through each drive and a few hours of 'cocks only' would be about the maximum in any one year.

Because a few superlative shots like Lord Walsingham were invited to the major driving days of many estates, a belief has grown up that the shooting men of that era enjoyed driven pheasants on six days of every week for months at a time, but that is far from true. In reality, if a landowner invited five of his friends to shoot his pheasants he could expect to be asked by each in return, and no more. Some modern syndicates are, in fact, little more than the formal pooling of resources which previous generations practised by tacit consent.

In the nature of things those tremendous days of rapid fire at driven pheasants, with bodies thumping on the ground like windfall apples in a gale, are remembered and discussed; but they are no more typical of the shooting of the time than the annual point-to-point is typical of riding.

It can now be seen that the decline of the partridge probably began in the mid-1920s; and it was being studied in 1932 by a forerunner of the Game Conservancy, then called the I.C.I. Game Research Station. At the time, this decline was thought to be no more than an unusually long sequence of poor seasons and it was not until after the Second World War, as I recall, that many shoots adopted the policy of breeding a few more pheasants to fill the gaps left by the shortage of partridges.

The change came almost imperceptibly during the 1950s. Everyone thought and hoped that the partridge would recover and we only bred more pheasants each year so that we should have something to shoot until that time came. It was with a distinct shock that we discovered that instead of having a partridge shoot with a few pheasants the latter had become the mainstay. In the south-east of England the balance swung in the late 1950s.

It might well be asked why we did not breed partridges instead of pheasants, and the answer is that many able men tried and failed. With the knowledge and equipment then available, it could not be done well enough; and in this connection it should be remembered that the difficulties are not confined to producing the eggs, hatching them and rearing the chicks. The real problem lies in keeping the partridges on the shoot after they have been released. Try as you would there was little to show for your pains, but if so much work and money were devoted to pheasants, particularly to hand-reared pheasants, there was almost certain to be a commensurate and predictable amount of shooting.

Some success was achieved with mallard, which are easy to rear but difficult to keep on the property, but the fact of the matter is that most lowland shoots now concentrate on pheasants for no better reason than that they are the only game birds which can be reared in large numbers and relied upon, more or less, to be present when they are wanted.

Pheasants have their virtues; they are beautiful to look at, excellent eating and are sometimes wildly exciting to shoot. As the dates can be arranged months in advance, things can be stage-managed so thoroughly that a tolerable shot who knows nothing of game or field-craft can hold his own in the front line. But that is about all. They do not deserve the aura of glory, the cachet, which attaches to them and they are not the epitome of excellence, superior to all else, and monarchs of the shooting field.

At the same time it would be wrong to belittle either pheasants or those who shoot them. They are not so easy that a newcomer can scintillate after no more preparation than buying an expensive gun and having a few lessons. Neither will an experienced wildfowler, pigeon shot or clay specialist necessarily do well without a period of acclimatisation; but all can learn in time to hold their own. The wealthy may enjoy sport on a larger scale but a short purse does not disqualify a resolute man.

There are those who maintain that although wild pheasants have merit, the hand-reared hardly come up to their standards; and in so doing they brand themselves either as poseurs or as woefully ill-informed. Admittedly, miserable little specimens trundling along a few feet above the ground are cannon fodder which should never draw fire; but those are simply bad pheasants. They may have been bred in the wild or in the rearing field or, let's face it, in both. It is true that hand-reared birds may be more difficult to flush at the beginning of their first season, but apart from that it is impossible to distinguish them from their wild cousins unless they are wearing rings or wing-tags. Their size, weight, strength, speed or ability to beat the guns is not dependent upon the place of their hatching but is controlled by much the same factors as affect the performance of any other athlete – for a pheasant is, or should be, an athlete in hard training.

One of the pheasant's greatest shortcomings is that it provides an unsporting shot when walked-up. When first flushed it is an easy target and a pale shadow of the soaring speed-merchant the same bird may become when driven. I write 'may become' rather than 'is' because early in the season on flat ground the average driven pheasant is not particularly difficult. In undulating country, strong pheasants helped by even a moderate wind will extend good shots; and in exceptional circumstances, which are so rare that many experienced men have never seen them, they will make all but the most skilful miss like novices.

It follows that driven pheasants are to be preferred but this poses a chain of problems for the manager, the greatest being the expense of beaters, the difficulty of arranging small days and, hence, of having a satisfying number of shooting days in the season if no other game is available. In the 1975–6 season, beaters cost £3·50 a day in my district; between twelve and eighteen are usually necessary to flush the birds, while from six to eight guns must be in front to intercept them. Such an assembly calls for an amount of organisation which is not justified by a few hours shooting; it must be all day or nothing, and that disturbs a lot of ground and kills many birds.

Everything is splendid the first time through; the next is less good, and the game will be wiped out or chivvied over the boundary if the same area is shot frequently. Things vary with the nature of the land, but about 300 acres are disturbed by one day of driven pheasants in my district and, no matter how many birds are present at the beginning of the season, there can only be two or three good days each year, plus something of 'cocks only', on any one piece of ground. If driven pheasants are to be shot once a fortnight from mid-November to the end of January, a big area of properly -keepered land is essential, and there must be an adequate number of beaters.

The much-discussed economy measure of arranging for the guns to take turns as beaters deserves comment. It works like a charm in the open, provided that discipline is maintained, but in a wood with anything approaching decent undergrowth it is essential that the men should put down their guns, arm themselves with proper sticks, lay about them with ordered vigour and penetrate first to the heart of every thicket and then right through to the other side. It is utterly and finally impossible for a man carrying a gun to do any such thing; and the usual stroll through the open spaces by a ragged line of men carrying a gun in one hand while flicking at twigs with a shooting stick held in the other, is a poor substitute. Such a parody of beating may flush a few pheasants by blundering into them but the majority will elude the human beings without even being seen.

Wild pheasants versus reared

This has been the subject of endless, inconclusive debate, largely because no two shoots are identical. I shall, therefore, only attempt to set out the general principles. It is foolish to attempt to rear pheasants before it is certain that they can be released into a tolerable environment which is reasonably free from predators. In an area so well suited to pheasants that the wild birds could produce an adequate shootable surplus, I should rear none. This is not the case in most ordinary mixed farmland, so the best long-term policy is to supplement the wild stock from the rearing field; and it is in striking the correct balance between the two that the manager's great problem lies.

The crux of the matter is not the number of shootable birds which the wild stock would produce if conditions remain as they are, but what that number would become if the keeper's work was confined to improving their lot. But no one can be certain that the wild birds will produce such-and-such a number, leaving the rearing field to provide so many to make up an acceptable bag. If the manager is working for himself alone he may gamble on his judgment, knowing that a mistake may ruin his shooting for one year; but if he is running the shoot for a syndicate he is more likely to play safe and rear rather more birds. And in so doing he defeats his own purpose.

That decision certainly ensures that an adequate number of pheasants will be available, although they will be rather more expensive, but, and this is the deadly blow, it also confines a single-handed keeper to the rearing field and the release pens for months. Then the beat-keeping is not done, predators thrive, the habitat goes downhill and the need for rearing on a large scale is perpetuated.

The proper solution is to give the beat-keeping precedence and to rear only as much as spare time permits; but if the members of a syndicate insist upon having something to shoot at once, the manager cannot take that course. All he can do in real life is to work in the direction of the ideal.

Two different approaches

On a modern, small shoot it is by no means unusual for the employer, the manager and the gamekeeper to be one and the same man, but let us imagine that a professional keeper is talking to his employer and that the target is a bag of 500 pheasants for the season. The keeper might well say: 'With six good shots in front I can almost guarantee a bag of 500 pheasants. We should rear 1500 chicks and that will require so-and-so equipment. Shooting twice through each drive and two days of "cocks only" would be about right. Now, about dates, Sir, . . .'

On these lines things can be forecast months ahead and the cost can be estimated with certainty; and if bad luck intervenes an extra shooting day will almost always make up the bag. This certainty and the limited liability appeals to some employers and to keepers who feel that their jobs and their tips depend upon the size of the bag. The method is not only the most reliable and predictable but it calls for the least exertion by the keeper: that it is also by far the most expensive concerns the employer rather than the keeper.

Another keeper might say: 'I cannot be certain of the outcome, but we might have more sport for less money if we left plenty of hens on the ground and reared about 200 chicks as an insurance. I could look after those in odd moments and spend my time killing vermin, chasing trespassers away and giving the wild pheasants a chance to breed in peace. If I could have a tractor now and again I could plant one of the game food mixtures to keep the bills for corn down. After an average summer we should have a bag of 500, more after a good one, but there would be next to nothing to shoot after a summer as bad as that of 1968.'

Those two imaginary conversations state the alternatives; each may be right and each may prove to be wrong on any one shoot. The one thing which is certain is that numerous predators or an unfavourable habitat will ensure the failure of either. When doubt exists, and the more a man knows of the subject the more likely he is to doubt, I suggest that the manager should enlist the help of the Advisory Service of the Game Conservancy.

The history and habits of pheasants

The more a man knows of the ways and tastes of game, the more successful he is likely to be. Some gardeners have 'green fingers' and there is an equivalent among gamekeepers which may be an instinctive skill but is more likely to come from experience, observation and thought. If all else is equal I am guided by my 'fingers', but they are never allowed to overcome logic.

Learned men tell us that pheasants are not natives of these islands but were first introduced by the Romans. Be that as it may, they have adapted themselves so thoroughly to life in our countryside that they are found almost everywhere except in vast tracts of heather or in coniferous forests. The key to their success is that, without being outstandingly good at anything, they are talented all-rounders. Pheasants can not only survive but flourish where the more specialised birds perish. By nature they are really birds of the reed beds and fresh water swamps, but although that is still the environment in which they thrive best, their diet is so varied and their adaptability is so great that they prosper wherever mixed agriculture is practised.

The variety of colouring is due to the importation of different species in relatively recent times. The Blacknecked, or Old English, pheasant is generally regarded as the oldest resident, with the Mongolian, Chinese and Manchurian the most recent arrivals. The very dark birds, which are now called Melanistic Mutations, are nothing more nor less than a natural mutation which occurred quite recently and multiplied.

All the varieties mentioned have equal merit in the shooting field; indeed their blood is so mixed by generations of cross-breeding in the wild that the vast majority are mongrels. I am quite content that this should be so, since I do not believe that any of the pure-bred varieties have any significant advantages or shortcomings, except that I discourage the melanistics. When first hatched, all melanistic chicks have a light-coloured patch on the 'elbow' of the wing at which the others may peck, and as soon as they have tasted blood those fluffy, endearing little babies start 'feather picking' like sadistic fiends. They will kill each other in dozens if given the chance.

Pure-bred pheasants can be obtained from some game farms; if compelled to specialise, I should choose Blacknecks because they tend to be the biggest and to lay the most eggs but, all things considered, healthy mongrels are inferior to none. The one thing not to do is to allow the so-called 'ornamental pheasants' loose in the woods. The Goldens, Silvers and so on are strikingly handsome but are no good for shooting and tend to be quarrelsome.

Although reed beds are the pheasants' favourite zone they are essentially birds of the fringes, forever moving from swamps to dry land, from cover to open spaces, from sunlight to shadow; and they usually travel on foot. The normal daily routine is to flutter down from roost at dawn, have a light meal and then sun-bathe, often on a gate or fence, while they preen their feathers and get warm and dry. Then they drift away from the roosting place in search of food.

If they have worked hard before hunger is satisfied, they settle down to have dust-baths and rest until well into the afternoon, but if they find enough food easily they go striding off in search of adventure, and that is just what a good keeper tries to avoid. All too often they either meet with disaster or stray over the boundary at this time. Later in the day they have another meal and make their way towards the roosting place at a pace which brings them home shortly before sunset.

It is worth noticing that, unless they are frightened, they seldom use their wings except when going up to roost and that they drift away from their home ground in the morning but towards it in the evening. This concept of ground which they regard as home is important. Pheasants are great wanderers, particularly those at the lower end of the pecking order, but straying is reduced if the keeper can plant the idea of a nice, safe Heaven on Earth in their minds. To this end, some arrange raised rails for sun-bathing, places for dust-baths, drinking troughs, grit and so on in a quiet, sunny clearing within a wood; and that spot provides a sanctuary.

When roosting in trees, pheasants choose a bough about an inch in diameter in a tree which gives shelter from the wind, but they also sleep on the ground. Because it exposes them to foxes this is not to be encouraged, but it is easy to make matters worse. A sleeping pheasant which is disturbed can fly into the night and it may land safely on the ground, but what happens then? They are not very good at settling on boughs, in fact the youngsters are downright bad, and it is doubtful if a bird which is flushed after dark will find a safe roosting place that night.

We tend to think that wheat is the staple food of pheasants because that is what men generally give them, but their diet is extremely varied. Wheat, barley, oats and maize are all eaten, together with all the grass seeds and most of the berries. Flies, insects, grubs, wireworms, leatherjackets, caterpillars and possibly beetles are among their favourite foods in summer, which is why they haunt dung heaps. Acorns and beechmast are among the staples and, as we have seen, it is said that they eat sweet chestnuts. Clover is on the menu, together with a multitude of cresses, the seeds of all the legumes and of most weeds. Curiously enough, they will always eat fresh meat and sometimes carrion. Some allowance must be made for curiosity and eccentricity and because a pheasant pecks at every plant in a garden it does not follow that any will be a popular food; but those loitering on the drive are almost certainly picking up the grit they must have to grind up their food.

It might be thought that this need for grit would be too well known to deserve comment, but it is not. At least, two grown men were distressed by the number of pheasants which assembled on a certain road and were killed by vehicles. The suggestion that a lorry-load of grit scattered in the woods would keep the birds at home was received coldly and they put wire netting along both sides of the road; and when the pheasants flew over it and were run down by cars both men commented on the stupidity of birds. Believe it or not, that is true!

Pheasants have a reputation for being bad mothers but it is doubtful if it is justified. Admittedly, those hatched in incubators and brought up in pens make a hash of rearing their first broods in the wild, but they have no memories of childhood to guide them. They seem to learn the ropes if given time and although rearing by hand ceased completely during the Second World War, the pheasants survived pretty well despite the merciless hunting inspired by small meat rations.

They do not pair off like pigeon or partridges; a cock gathers a harem of up to six hens but he usually plays no part in hatching the eggs or looking after the chicks. The nest is always on the ground and almost invariably on the edge of something. A grassy bank with some cover overhead is often chosen, perhaps under a hedge, around a pond, on an embankment of any kind or in the grass verge of a road. Grass-covered clearings and rides in woods are popular but nests are seldom found more than 20 yards into a stand of trees. As with partridges, many nests are destroyed by grass-cutting machines because they are so often made in growing hay or silage. Marking such nests and mowing round them, rather than over, was a common practice in the old days and is still rewarding.

The first eggs are laid early in April, or at the end of March if spring comes early, and about two eggs are laid in the course of every three days thereafter. In the wild, the hen lays a clutch of about twelve eggs and then begins incubation. If all goes well, the chicks hatch twenty-four days later and, like baby chickens, are fully mobile as soon as they are dry. The mother leads the brood away from the nest, never to return, and the chicks follow her; they pick up their own food from the outset but they depend upon her for warmth and protection from the weather.

Their diet at first is much the same as that of baby chickens and there is usually plenty of food available, but they are terribly vulnerable to bad weather. When brooded by the hen they are warm and dry but cannot find food; if they become wet or cold when searching for it, survival depends upon reaching the warmth of their mother's feathers quickly.

Should the hen lose her eggs she will probably try again, and young chicks have been seen early in October, though how many previous failures had occurred can only be guessed. Although she will only lay about a dozen eggs in the ordinary way before going broody, if the eggs are removed every day she may continue to lay right into July and produce over forty eggs in all. Some of the reasons for the resilience of the species can be seen here: pheasants lay so freely over such a large part of the year that some good weather is almost certain, and losses may be replaced.

How the keeper can help wild birds

Now let's go back and study what forms disaster may take, because that shows how the keeper can help. The eggs are in danger from badgers, rats, hedgehogs, foxes, stoats, all the crow tribe, moorhens, human beings and dogs, from the time the first is laid until they hatch. By far the best defence is to kill the predators, failing which the nests should be difficult to find or apparently dangerous to approach. A circle of Renardine or creosote on the ground about 3 yards from the nest will discourage varmints.

Sitting hens are constantly taken by cats and foxes, particularly in the twenty-four hours prior to hatching. A sitting hen has less scent than is normal but she seems to lose this asset right at the end of incubation; the same is true of partridges and mallard.

The chicks are easy meals for cats, rats, stoats, mink and weasels. Jays and magpies take a toll, as do crows and some of the gulls. By the age of nine weeks the poults can fly a short distance but they are not good at settling in trees and are often chased to exhaustion by dogs and foxes. It is not until late in September that young pheasants fly well enough to beat their enemies by the strength of their wings, and until that time comes their main protection is concealment. The wonder is that enough survive to maintain the species and it is reasonable to suppose that Nature enabled pheasants, duck and moorhens to lay so many eggs to offset the heavy casualties among the young.

That is the beacon which lights the keeper's road to success. The data is none too reliable but it is probable that the average pheasant nesting in the wild has no more than two offspring alive in the following October, yet the species survives. If man could reduce the losses by even one third, no further action would be necessary because the pheasants themselves would give a sizeable surplus to shoot. Their best chance would occur in some serene cloister or sheltered, walled garden; so that is the keeper's target.

Unhappily, picnic parties, organised walkers and townsfolk in general do not realise that their blaring radio sets, yelling children and wildly excited dogs are the negation of all the keeper most desires. In their own eyes they are harming nothing and may even think that they have a right to be where they are; but if they are asked to go away they generally grumble and do so. If, however, they are shown a few distant pheasants or a nest with eggs (dummies suffice) they see the point and become allies.

Conclusions

I have tried to give a balanced and impartial description of the attractions and limitations of pheasants so that readers may either go ahead with their eyes open or steer clear for sound reasons. They are not reserved for the wealthy, beyond the skill of non-experts, a mark of social standing or a proof of decadence, but simply the best available way of ensuring that some shooting will be possible.

When Fate is kind, the pigeon may provide hectic sport for the cost of cartridges alone. Wildfowl can be every bit as satisfying as pheasants, and the same is true of snipe and woodcock; but they are all fickle charmers and only the pheasants are dependable.

There is a deep-rooted belief that rearing pheasants by hand is essential for the enjoyment of good, lowland shooting in modern England, but a great deal depends upon the definition of good shooting. If seven men are to assemble at 9.30 am on every alternate Saturday from mid-November to the end of January, be escorted to stands from which they will fire an entertaining number of cartridges and be reasonably certain of seeing a specified number of birds in the bag at 3.30 pm, then pheasants there must be; but they need not necessarily be hand-reared. On good land the skilled management of wild game may well suffice, and it should be far less expensive.

On ground which is fundamentally incapable of holding an adequate stock of game or, which is far more common, has not yet been worked up to its full potential, some rearing is essential; and it should be accepted that the greater the proportion of reared birds the higher the cost of each pheasant in the bag will be.

Even so, it is still possible to have good sport at not great expense if the men are content with small bags obtained with the help of intelligent field-craft – and that presupposes a small group, all of whom know what they are about. If these men are prepared to forego shooting when prospects are poor but to grasp opportunities when Lady Luck smiles, they may have memorable shooting very cheaply.

13

What Good Management Can Achieve Without Rearing

(as demonstrated by Count Louis Karolyi)

To shoot over 40,000 acres, employ twenty-five keepers, have annual bags in the region of 30,000 head of game and show a profit almost every year might be regarded as no more than a vision of Nirvana; but it was done by Count Louis Karolyi during the fifty years prior to 1939. Moreover, this was achieved on land which was rather unfavourable to game and without any substantial number of coverts until he planted them.

No game was reared by hand and these results were obtained by a blend of sound agricultural and shooting policies pursued over a long period of time. This was not an instance of an enthusiastic millionaire pouring wealth into his shooting, for the sale of the game covered the costs; neither were the interests of the farms sacrificed to those of sport.

The change in the value of money conceals the fact that the staff were distinctly well paid, and they were also encouraged by a commission of 10% on the value of the bag, in addition to a bounty paid for the vermin destroyed. I have emphasised that Count Louis Karolyi's shooting was on well-farmed land not especially suitable for game and that the staff was well paid, because the estate was centred upon Totmegyer, which was at one time within the Austro-Hungarian Empire.

It is easy for detractors to paint a picture of grossly ill-used workers by stating the wages they received while withholding the information that those, now derisory, sums sufficed for a very satisfactory way of life at the time. Also, English opinions of that part of the world have been coloured by the writers of fiction and light opera. In reality the country has little in common with Ruritania, life there is neither exceptionally drab nor outstandingly gay, the population does not contain a high proportion of eccentric characters and the influence of Count Dracula is negligible. Neither was a starving and unwilling peasantry driven into the shooting field in Count Louis' time. The estate survives under the name of Palarikovo. The shooting is now run as an attraction for tourists and the bags are still enormous; those who wish to explore the possibilities of a shooting holiday in that region should enquire at the Czech tourist office, Cedok, 45 Oxford Street, London.

I do not suggest that it is either possible or particularly desirable to re-create such shooting in modern England, but as such brilliant results were achieved without cost in conditions not markedly superior to our own, it is at least worth devoting a little time to the study of the methods used.

The annual bags were of the order of 10,000 pheasants and the same number of partridges and hares. 20,000 acres were owned and the shooting over another 20,000 was rented. This is a total of 62 square miles, or rather less than a rectangle 8 miles each way, which is big by English standards, but not phenomenal, and the keepers were not unusually thick on the ground. The average is 1600 acres for each man, but as the head-keeper and his deputy were probably confined to supervision, the beat-keepers and assistants may have had about 1700 acres each. It is accepted in England that 1100 acres can keep a man busy and that 2000 is the top limit, so with no rearing to distract them it can be assumed that the keepers were neither pampered nor over-worked.

About 10,000 head of vermin were destroyed each year, and this works out at no more than 1·33 head per keeper on each working day, which is a low score by modern English standards. The protracted war waged over a large area must have thinned down the undesirables to a fraction of the density we endure. (The isolated small shoot in this country is bedevilled by predators which come over the boundary from un-keepered land as fast as the resident rogues are destroyed. Only if their numbers are kept low can game breed freely in the wild, and this is hardly possible unless the neighbours join in the hunt.)

The soil is heavy clay, which has disadvantages for game, but the continuous wind of springtime dries the crops quickly after rain and reduces the losses caused by damp among young birds. The chief crop in the main shooting area was sugar beet, which is popular with game, but the method of working the land contributed to their welfare. There was almost no mechanical equipment, and slow-moving oxen did the heavy work and allowed the game to get out of the way; also many oxen mean much dung and swarms of insects. Research on the estate showed that insects and grubs were the principal food of pheasants during the spring and summer and that as the stock of game increased over the years so did the number of insects harmful to agriculture decline. This, of course, was before the days of spraying. The area was thinly populated and there was none of the coming and going which motor cars and better roads have made possible.

So much for the advantages of Count Louis' estate, but his game had to endure the severe Continental winters without the shelter and food provided by our woods. It is at least arguable that pheasants are more likely to thrive in average English conditions.

As both the game and the farms were worked up, the annual bag of pheasants and partridges levelled off at about 10,000 head of each, with the same number of hares. The break-even point financially was a total of 15,000 head. The intention was to leave a breeding stock of between 25,000 and 40,000 head upon the ground so, assuming that partridges and pheasants were left in equal numbers, between one-and-a-quarter and two birds were left for every one which was shot; which is a very high proportion by English standards.

The game was fed upon massive quantities of low-grade wheat and maize – and 'massive' is indeed the word: the breeding stock alone would gobble up something between $1\frac{1}{2}$ and $2\frac{1}{2}$ tons of food on every single day. Yet the policy was successful. All criticism is silenced by the fact that the shoot paid for itself; that is the inescapable proof of the pudding.

During two years in which pheasants were exceptionally numerous, both cocks and hens were shot, but in normal times it was 'cocks only' throughout the season. That is the only major difference I have been able to discover between Count Louis' methods and those in common English practice; but he set great store by a method of avoiding damage to nests which was once general in this country, although it has now fallen into disuse. Before a field of hay is cut, a line of men carrying a light rope strung between them walk abreast through the area, allowing the rope to brush the top of the grass. This flushes sitting birds, the nest is found and marked with a stick and the mower leaves an uncut patch around it. If the hen returns all is well, but it was the custom in this country to hatch the eggs at home if she deserted. It is very rewarding with pheasants but is so wasteful of man-power that it may be wiser to put the effort into the rearing field nowadays.

What then were the secrets of Count Louis' success? A clue exists in the competition which was seen when any vacancy occurred on the staff. The keepers thought well of their jobs and, presumably, worked hard to retain them. An adequate number of trained and willing gamekeepers was undoubtedly one vital factor. The size of the shoot made things easier by reducing the harmful effects of un-preserved land nearby, but there is no record of Count Louis doing anything which would be considered abnormal in England. I therefore conclude that his outstanding results sprang from the brilliant execution of a sound plan over a long period of time, rather than from black magic or unique methods. Phrased differently, this might amount to: 'His plan was no better than ours, but he was very, very good at putting it into effect.'

The need for good staff-work can be seen when it is known that an ordinary major day of shooting at Totmegyer involved 400 beaters, controlled by twenty-five keepers, and a total of thirty-two ox carts to carry game and baggage. A big day in November would yield something like 1000 cock pheasants, 500 partridges and 1500 hares; but the peak of the season was reached in December with three major hare shoots and three big days with pheasants. The latter employed 400 beaters on a total of twenty-four drives.

The best year was 1933, when the total amounted to 11,740 hares, 10,723 pheasants, 14,989 partridges, 111 rabbits and 53 'various'; but the record bag of pheasants was made in 1909, when eight guns shot 6125 pheasants and 223 'various' in one day. There are many small but well-run shoots in this country which do not shoot 6125 pheasants in ten years, during which time a great deal of money is spent; and that is a yardstick by which the standard of management at Totmegyer can be measured, for not one bird was reared and those pheasants did not cost a penny piece between them.

It might well be asked how the eight men who made the record bag could withstand so much firing, carry so many cartridges and retrieve so much game. The answer is that each man who was shooting had a total of eleven attendants, made up of loaders, cartridge carriers, pickers-up and game transporters. Seven of the men had three guns each and one had two guns. At the end of each drive the call, 'Guns this way, please', should have put almost a hundred men into brisk, ordered motion!

Two guns of the orthodox, double-barrelled, design are essential for shooting of this intensity but a third adds little, if anything, to the rate of fire. The true purpose of the third is to ensure that two will still be available even if one breaks down, and in the ordinary way it is most unusual for a mechanical defect to appear in a gun of medium or better quality. But the shooting at Totmegyer was far from ordinary. The guns were often heated by protracted rapid fire and then surprising things may happen to any which have been oiled too generously: disconcerting things such as fore-ends coming adrift or both barrels going off at once are by no means uncommon. Then a skilled gunmaker is needed quickly, and such was the standard of the preparations that a fully-equipped and capable repairer was waiting in the house to put things right.

I understand that Count Louis Karolyi has described his shooting in a book called *Waidwerk Ohne Gleichen*, but I have failed to find any English translation. Some of the information in this chapter was given to me by the

Count Louis Karolyi who is a nephew of the Count Louis to whom constant reference has been made. The latter took over the management of the shooting at the age of 17, when his father became the Foreign Minister of the Austro-Hungarian Empire. Both Mr I. G. Bellak and Mr Pal Mariassy have also given help, and to all of them I am more than grateful.

14
Rearing on the Smallest Scale

Even though you set your face firmly against rearing birds by hand, it will not be long before an apron full of pheasant chicks is thrust into your hands with the news that their mother was killed by a mowing machine. Compassion will overwhelm logic and you may feel rather helpless, but you can be reasonably certain of a happy ending if you follow these instructions.

Carry the chicks home in a warm, ventilated cardboard box, if possible; give them water to drink and a meal of chopped up hardboiled egg and keep them warm, not hot, by some means such as a hot water bottle or a suitably screened electric light bulb. Locate a broody hen which is sitting on eggs and, at midnight, remove the eggs without disturbing her and substitute the chicks. It does not always work but she will usually adopt them quite happily.

If you establish the hen in an ordinary coop with a run 8 or 10 feet long and 6 feet wide on the lawn she will do the mothering. You must provide water, food and grit, shut the brood up in the coop at night when the chicks are small and move the run onto fresh ground every day.

The staple food should be pheasant or turkey crumbs, preferably with a coccidiostat to build up resistance to coccidiosis, but ordinary chick crumbs will do at a pinch. These crumbs can be obtained from corn merchants and are advertised in the shooting magazines. It is important that the right sized crumbs and grit should be given for each age group – the instructions are printed on the package – and the old hen eats whatever the chicks are given.

The best hens for this purpose are the self-reliant specimens of indeterminate breeding which range about stackyards and forage in the fields. Rhode Island Reds or others of a similar disposition will serve and the ordinary, common bantam makes an excellent mother. A hen can look after about fifteen pheasant chicks and a bantam about six.

This method is the basis of the most practical way of rearing up to about six broods of pheasants. The method is the same whether the broody hen hatches some pheasants' eggs or is given day-old chicks bought from a game farm. The upper limit is about six broods, partly because broody hens are hard to find nowadays and partly because other methods are less time-consuming for larger numbers. The one thing not to do is to allow game to use the same ground as ordinary poultry; it seems that their natural immunities to disease are different and that the quickest way to start an epidemic among both is to mix them.

To return to the broody with the orphans: the best way to establish these pheasants in the wild is to allow the old hen to wander about with her part-grown brood and to let them overflow into the fields as they wish. The

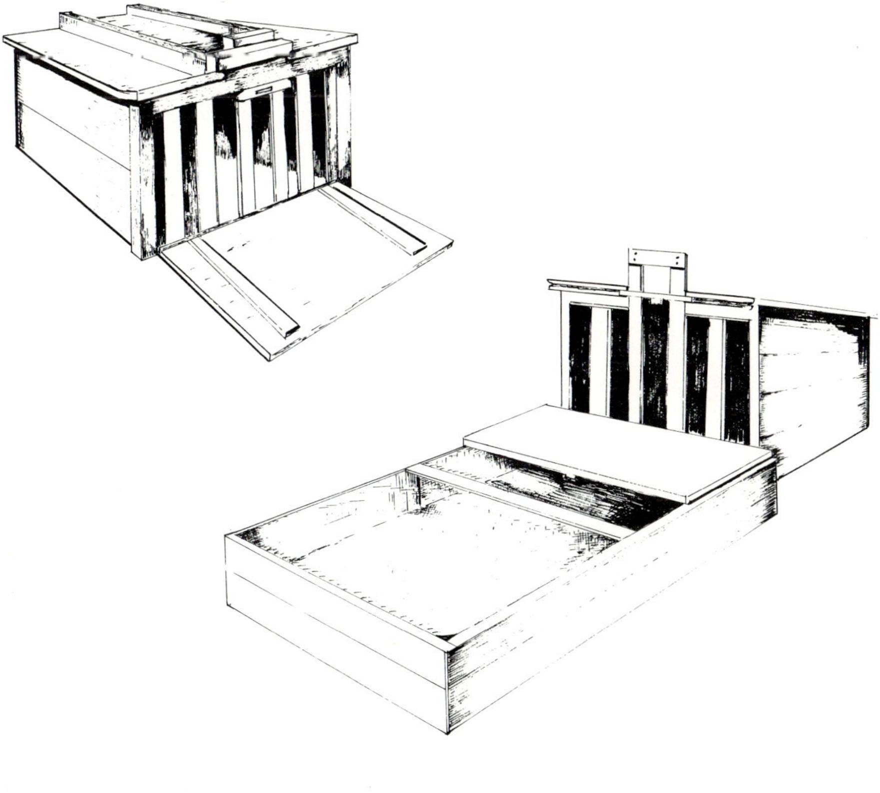

COOPS CAN BE MADE AT HOME, WITH OR WITHOUT RUNS

most effective action you can take is to do nothing. If they must be transferred to the woods, the old hen should be taken as well and established in a fox-proof pen of some kind. She will anchor her own brood and that community will help to anchor any motherless poults reared by modern methods.

Wild ducklings

Orphaned mallard can be reared under a broody hen in the same way if they are given the food of domestic ducklings, but they should not be allowed to swim before they are well feathered. A duck has oil on her feathers which waterproofs her ducklings' fluff; but a hen is without this oil, so when her ducklings swim they get cold and die. Resist the temptation to rear ducklings with the hen in a coop but without a run: they will ruin your garden by dibbling and paddling everywhere.

Partridge chicks

Partridges reared by an ordinary bantam can be a great success. Start them in a coop with a run, but then allow the bantam into the run. Later, prop up one corner of the run so that the brood can wander about and eventually find their way into the fields. The bantam learns to fly and partridges have more flock discipline than pheasants. The sight of the bantam flying back to the garden at the head of her covey on a December afternoon is well worth watching.

Rearing small numbers without broody hens

If no broody hens can be found, one of the electric heaters used by game farmers and pig rearers will keep the chicks warm, and an ordinary electric light bulb screened by a tin to produce twilight has served the purpose. As long as they are healthy, clean, warm and well fed, both chicks and ducklings are astonishingly robust; but without an old hen you must do the mothering, and motherless pheasant chicks are clueless beyond belief. You will probably have to tap the food tray to induce hungry chicks to eat and you will certainly have to shoo them in out of the rain.

15

Rearing Pheasants

'Rearing' is a term which covers the whole process of increasing the number of young pheasants by means other than allowing the old birds to breed in the wild as Nature intended. It embraces obtaining eggs by confining mature birds in pens, hatching them in incubators or under broody hens and bringing up the chicks, either with domestic hens as foster mothers or under artificial heat of some kind. Strictly speaking, rearing ends when the part-grown chicks, now called poults, are transferred to the woods and 'released', 'releasing' being considered a separate phase, but 'rearing' is often used loosely to include both phases. A newcomer may be bewildered by the innumerable alternatives, but things will be clearer once the limitations of each have been stated; in practice, the correct solution is usually evident once the problem has been stripped of non-essentials.

The different systems of rearing

THE MOVABLE PEN METHOD Chapter 14 describes how an ordinary domestic hen can bring up a brood of young pheasants: the hen is in a coop and the chicks have access to a run which confines them. That, in essence, is The Movable Pen Method. If you obtain any number of foster-mothers, establish each in a coop with a run and range them across a field of grass, on paper at least, a commensurate number of pheasants can be reared to the age of six or eight weeks. Good birds are produced in this way and casualties are few. The snags are that broody hens are difficult to obtain, the method calls for many man-hours of work and the capital cost of the equipment is high. The economic limit is about ninety or a hundred chicks.

THE OPEN FIELD SYSTEM If we put the domestic hen in a coop as before but allowed the chicks to come and go without the restraint of a run, we should have the essentials of The Open Field System which was used to rear thousands of pheasants every year in the old days. Hundreds of coops, each containing a hen with a brood, were ranged across vast fields, while predators from miles around rallied to the feast. The sole merit of the system is that it demands the minimum outlay of capital, but it is grossly inefficient and it works all hands to exhaustion.

FORDINGBRIDGE-TYPE EQUIPMENT This allows one man to rear five times as many birds as he can by the older methods, but it calls for a higher standard of management. We are now dealing with groups of chicks each a hundred strong and it must be accepted that the larger the group the higher must be the standard of hygiene, feeding, the control of temperatures and of management in general.

I should choose Fordingbridge equipment for any number of pheasants between one hundred and, say, sixteen hundred; but for larger numbers I should hesitate. In certain circumstances I should gamble on large-scale units, knowing that I was taking risks. No matter how many chicks were the target I should bear in mind that the Rupert brooder in Fordingbridge-type pens can overcome difficulties which have no other solution.

Before describing the construction and use of Fordingbridge-type equipment, let's break the business of rearing into its components, because you can buy in at any level you choose and opt out of all the stages before that point.

The stages in rearing

EGGS ARE ESSENTIAL You can either: (a) buy eggs from a game farm, or (b) pen mature birds together in the ratio of one cock to every six hens to produce your own, in which case (c) you might sell your surplus eggs to offset expenses.

HATCHING THE EGGS All large-scale hatching is done in incubators nowadays. The alternatives are: (a) to buy day-old chicks from a game farm, when the price is always approximately double that of eggs; (b) to have your own eggs hatched at a central hatching station, or (c) to hatch eggs in your own incubators with, or without, setting up in business on your own account as a central hatching station and selling the surplus chicks.

REARING DAY OLD CHICKS TO THE AGE OF FOUR WEEKS Unless chicks are reared under domestic hens, they must have artificial heat of some kind for the first four weeks of their lives. This heat may come from electricity, cylinders of gas or paraffin burners. You can either: (a) arrange the supply of heat and rear the chicks yourself, or (b) let a game farm do the work and buy chicks four weeks old, by which time they will have out-grown the need for artificial heat. Naturally, a bird four weeks old costs more than a day-old, but we will return to the economics later.

REARING FROM FOUR WEEKS OLD TO THE AGE OF RELEASE Whether you buy chicks four weeks old or rear them yourself, they must still be kept in pens with adequate protection from the weather until they are big enough to be established in release pens in the woods when between six and eight weeks old. During this period they need no artificial heat, except perhaps on the occasional frosty night, and you can either: (a) do this work yourself, or (b) buy poults old enough to release at once at a substantially higher price than four-week-olds.

RELEASING This is where the honourable options run out. You can escape any of the earlier tasks by buying from a game farm, but from the age of six or eight weeks onwards the work must be done on the shoot. Releasing will be described later; suffice to say at this stage that it demands more skill but is less laborious than rearing up to the age of release.

The dishonourable alternative is not to release the pheasants but either to keep them in pens or arrange for them to be delivered in baskets on the morning of the shoot. Then attempts are made to shoo them over the guns. This method is called 'Instant Pheasants' and I want no part in it for several reasons, any one of which is sufficient by itself: humanity, they cannot fly well enough to be worth shooting, and even if they could they are far too expensive.

The case for and against buying at some stage from a game farm

Things may be less confusing if I say at once that a beginner would be wise to buy day-old chicks from a game farm and do the rest himself. As experience brings knowledge and the ability to make sound plans he should aim at greater numbers, produce his own eggs and consider hatching them in his own incubators; but he should refrain from doing so until he is strong enough to start a central hatching station of his own. Most of the reasons for this decision will emerge without detailed explanations.

The game farmer's case runs much as follows. 'I am a skilled professional working on a large scale with enough capital invested to make use of all that science and mechanical appliances can offer. As such, I have the advantages of buying in bulk, keeping the plant running and employing skilled men at all stages. It is better that I should supply the birds for fifty shoots rather than have fifty amateurs blundering about with inadequate equipment and less skill. So let me rear the pheasants and let the amateurs do something useful, such as killing predators and improving the habitat, with the time they save through buying from me.'

The amateur keeper or manager of a shoot replies as follows. 'The game farmer is not invincible; if he can make a living I can learn to do what he does and save the profit he makes. I must acquire sufficient knowledge and skill but I have built-in advantages; all my overhead charges are lower, I pay nothing for wages, NHI, rent or rates, and, with the exception of the incubators, all the equipment can be made at home with a saving of half of the capital costs.'

Either the game farmer or the amateur may be proved right by subsequent events, and some amateurs sell their surplus to game farms and still make a profit on the deal, but not all the facts have yet been stated. Leaving hatching aside, all the work of rearing is essentially simple; once the pens have been set up correctly the teenaged daughter of the house can rear excellent pheasants, but that is not enough. Producing good physical specimens is easy, but producing them in large numbers at a sane price is quite another matter. If that is to be achieved, expensive plant must not be allowed to stand idle but must be driven hard without allowing standards to fall – and that involves drawing up a sound programme and sticking to it come what may. If the watchword is 'The Best Is Only Just Good Enough And The Timetable Is Sacred' the rearing of first-class pheasants cheaply presents no serious difficulty, but any toleration of the second-best or failure to maintain the timetable will be reflected in the quality and in the cost of the birds.

It is in the maintenance of high standards, of taut discipline and clockwork regularity that the professional excels most amateurs. If the supply of electricity fails, his emergency action is taken without panic or delay, he does not breed from poor stock, his family donkey does not get into the rearing field and knock pens down and never, never does he lose a day through not having things ready in time.

The most rewarding resolution a prospective rearer could make would be, 'I will have every single thing, from breeding pens to spare fuses, from release pens to medicines, food for the chicks and typed copies of the rearing programme ready by 1 April.' Then the inevitable crises will produce no more than moderate chaos.

Making Fordingbridge-type rearing equipment at home

Author's Note I dislike the mixture of metric and Imperial measurements which follows and it is only tolerated as a guard against mistakes. A difficulty arises because the timber trade has adopted the metric system: one of the new standard lengths is 3 metres, which is a shade less than 10 feet, so netting panels are now made 3 metres long instead of the 10 feet they were, and it follows that the pens must be 6 × 3 metres or 6 × 6 metres.

That is straightforward and we can cope with it, but two booby traps are lying in wait. The first is that although we pay lip service to metric dimensions we still think in inches and our tools are made for working in inches. Familiarity has taught us how strong a piece of 2 × 2 inches softwood is and which chisel should be used for cutting a mortice in the middle of it. Perhaps we shall be equally at home with metric units in ten years time but in the present transition stage we most certainly are not, and mistakes are being made as never before. The professionals who build houses have coined the phrase 'discrepancies in the measurements' as a euphemism; it should not be needed but habits die so hard that it is frequently heard.

The second route to chaos lies concealed in the fact that for small dimensions and quick reckoning the timber trade converts at the rate of 25 millimetres to the inch – and every schoolboy knows that the correct rate is 25·4.

There is no alternative to buying timber in metric units, but when the reader is following the working drawings in this book he can choose either metric or Imperial units. He will not go wrong as long as he sticks to one or the other, but he should not mix them.

All that is good about this equipment comes from what we now call the Game Conservancy, but which was known as the Eley Game Advisory Station when I and many others cribbed the design (with the consent and blessing of that organisation) and made the pieces at home. It is possible to buy this equipment complete in all particulars but any home-carpenter or handyman can make it. If he uses first-class timber bought at the market price he will save about half the cost of the ready-made; but if he searches the woods for wind-blown larch and fir, and cuts the logs to size with a circular saw, he need only buy plywood, wire netting and small stores.

Over the years, I have made many small alterations to the original design in the light of personal experience; so have many others and we all maintain that our own variations are the one, true path; but all credit should be given to Fordingbridge and any shortcomings should be attributed to those who have tampered.

The working drawings in this book will enable non-experts to make efficient equipment without wasting materials or labour. They are based on the designs I have evolved when trying to devise equipment which was effective, convenient to use, easy to make and a sound financial investment. This last point is the big problem; it is easy but unsatisfactory to make a shanty which falls to bits in six months, and a cabinet-maker's job built of solid oak would endure but would be too expensive to pay its way. I believe that the economical balance between the capital cost and the length of life has now been found, but I may be wrong in this because I do not yet know how long it will last. Some equipment made fourteen years ago is still going strong, while some built to a lower standard has perished. There are those who affirm that softwood will not last more than a year or two when exposed to the weather with no more help than an annual coat of creosote. When I reply that in my experience it is serviceable after fourteen rearing seasons of hard usage I am seldom believed.

To forestall criticism: netting panels 4 feet high are substantially cheaper but compel the rearer to stoop unbearably. Panels made with only three uprights are too frail. I made one brooder house of thinner plywood on a lighter frame but it lacks rigidity. It has served for four years, and may be an economy, but I prefer the thicker plywood on a stouter frame because it has a trouble-free life of over fifteen years.

One lesson taught by experience is that equipment should be uniform, interchangeable and reasonably well made. If you choose panels 5 feet high they should all be of that height, not an untidy mixture a few inches on either side of that measurement, and none should be shaped like the ace of diamonds. It is impossible to produce a labour-saving arrangement with equipment which is either badly made or is a series of improvisations; above all, avoid buying a variety of second-hand equipment cheaply at sales – in the long run it is likely to prove the most expensive of all.

When the last of the poults have left the rearing field all the equipment should be cleaned, dismantled, disinfected, given a coat of creosote and stored under cover until it is needed again. The insides of my brooder houses are not treated with creosote because it is said that heated creosote is bad for chicks.

Concerning materials: all the softwood timber is 'as sawn', not planed, although I smooth a few pieces to ensure a close fit. All the plywood is of 'exterior' quality, not 'marine' quality. Hardboard $\frac{1}{8}$ inch thick has been used in place of plywood with some success, but I have insufficient experience of this variation to form an opinion.

Some home-carpenters waste materials either by buying more than is essential or by cutting it in such a way that an unnecessary proportion must be discarded. These twin dangers will be avoided if the cutting diagrams in this book are studied. I strongly recommend that the professional carpenter's routine of 'Measure, mark, measure again and then cut' be adopted.

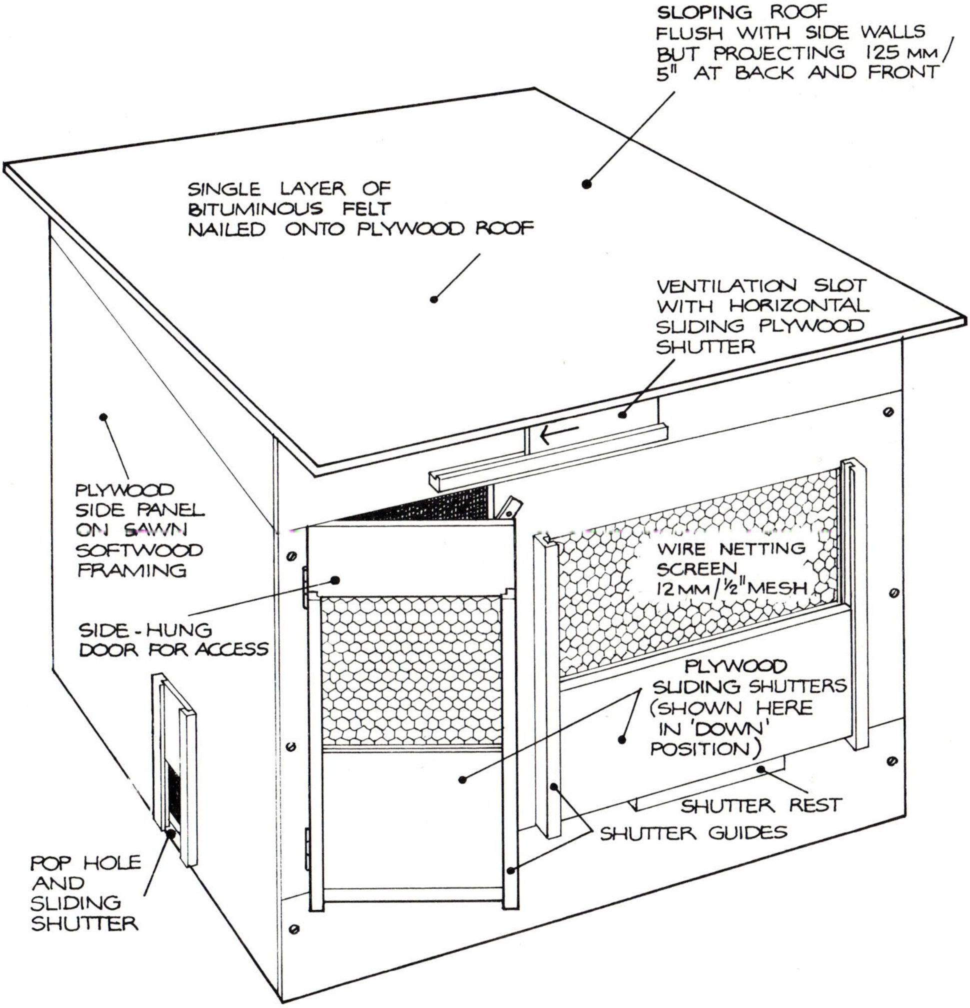

VIEW OF COMPLETED BROODER HOUSE
SHOWING FRONT, LEFT HAND SIDE AND ROOF

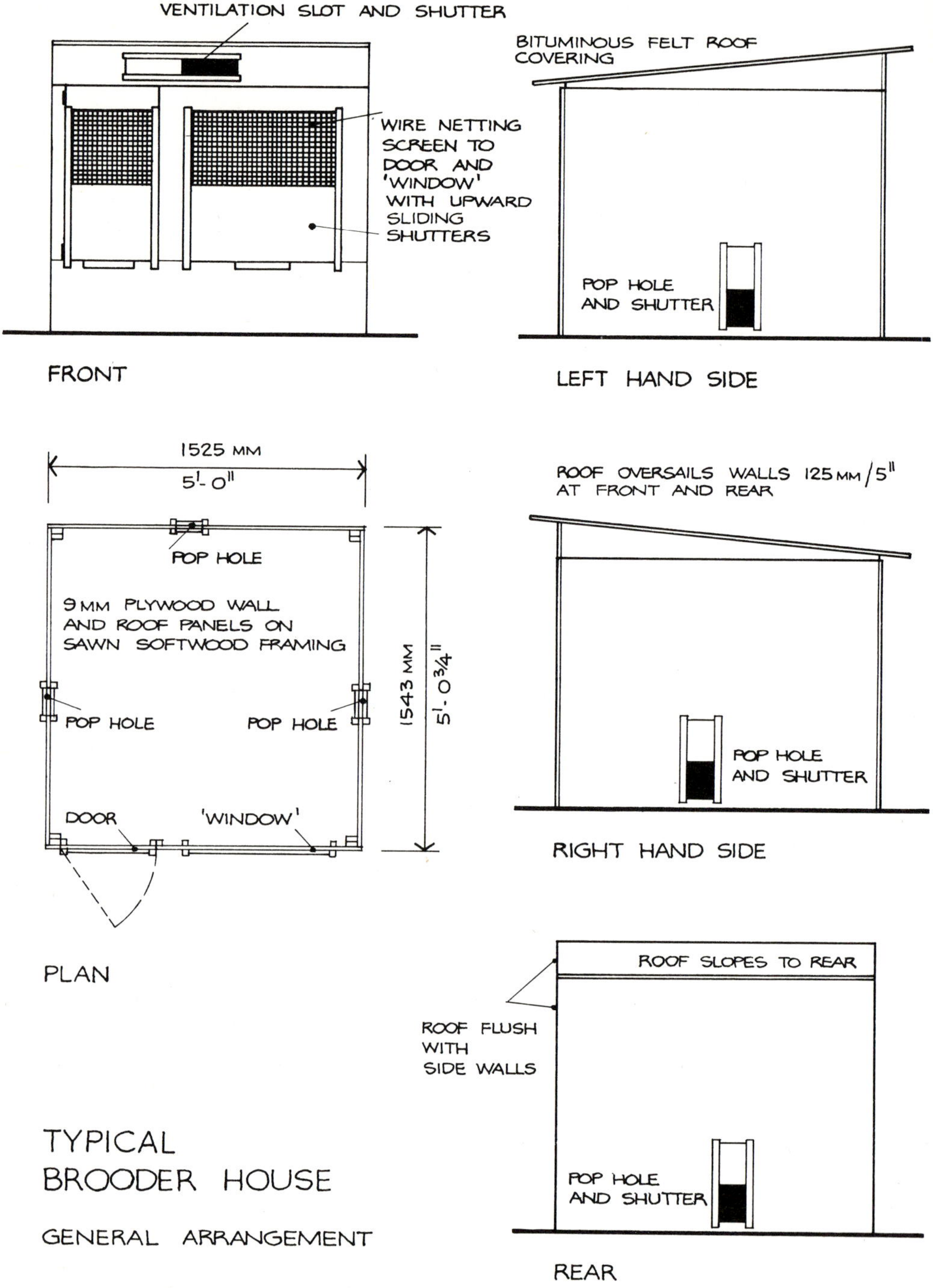

TYPICAL
BROODER HOUSE

GENERAL ARRANGEMENT

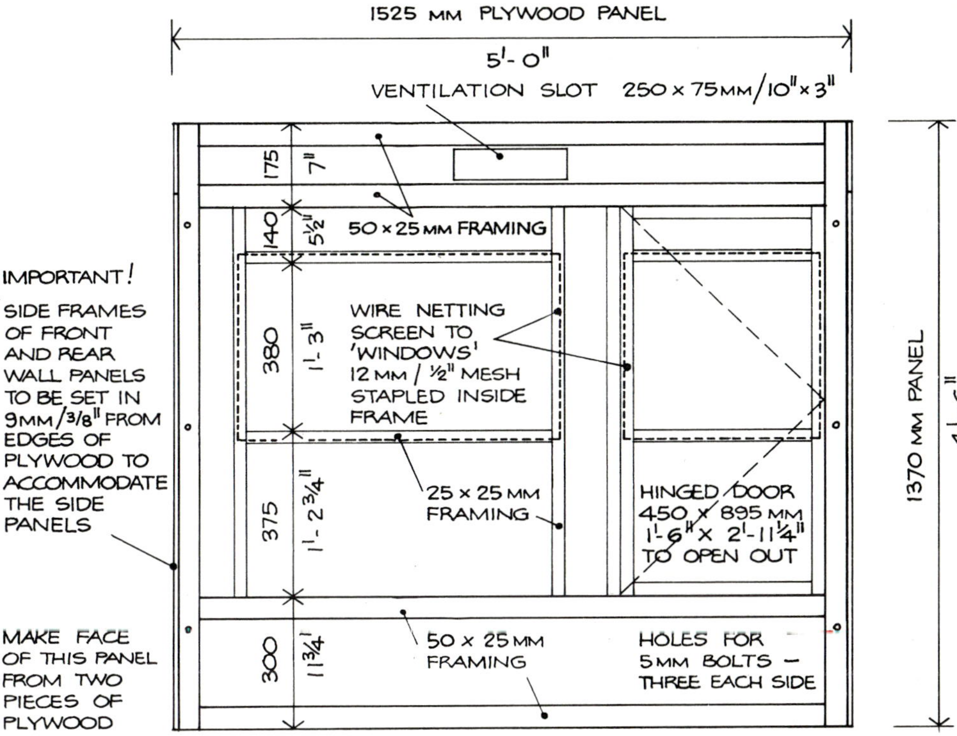

VIEW OF THE INSIDE OF THE FRONT WALL PANEL, SHOWING FRAME.

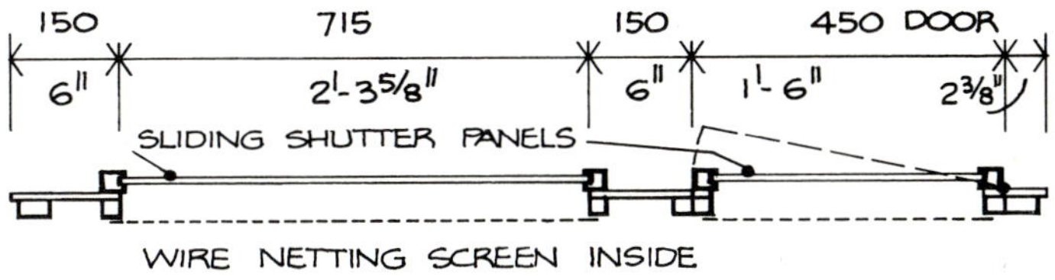

PLAN THROUGH FRONT WALL PANEL, AT 'WINDOW' LEVEL.

NOTE:
SOME OF THE METRIC DIMENSIONS SHOWN ON THESE PANEL DETAILS ARE NOT THE EXACT EQUIVALENT OF THE CORRESPONDING IMPERIAL SIZES, BECAUSE THEY HAVE BEEN 'ROUNDED OFF.'
THEREFORE, PANELS SHOULD BE MADE UP USING EITHER ALL OF THE METRIC OR ALL OF THE IMPERIAL DIMENSIONS.

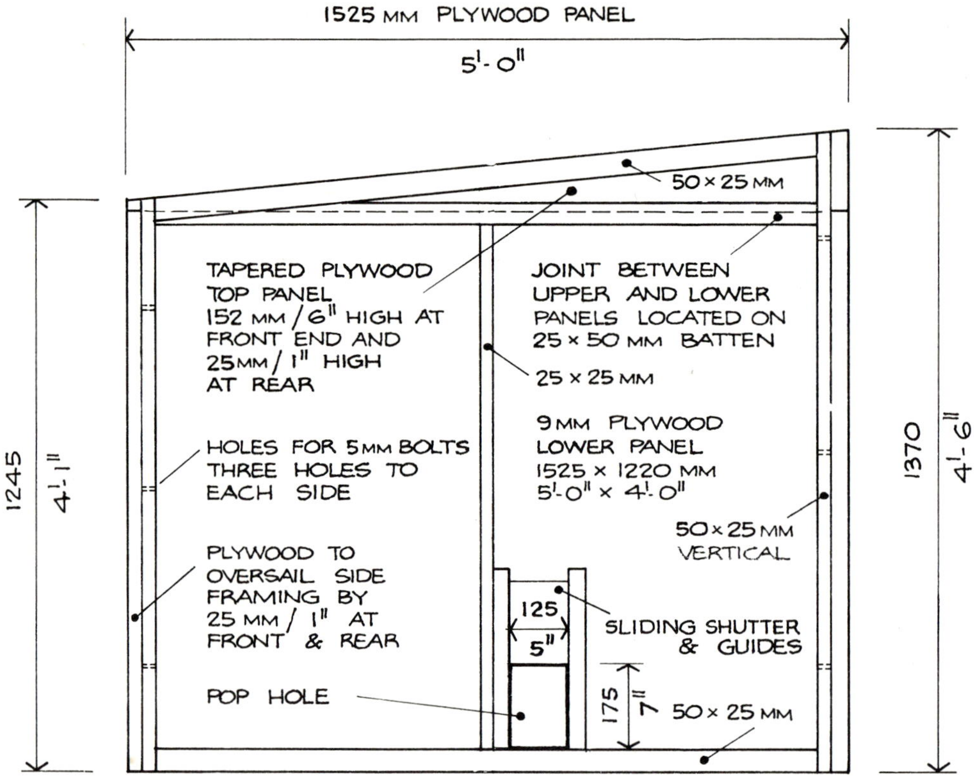

VIEW OF INSIDE OF RIGHT HAND SIDE PANEL

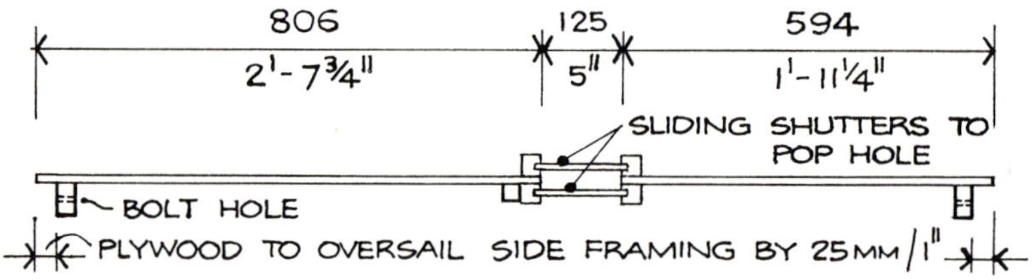

PLAN THROUGH SIDE PANEL AT POP HOLE LEVEL

NOTE:
DETAILS OF LEFT HAND SIDE PANEL ARE AS SHOWN ABOVE BUT IN REVERSE.

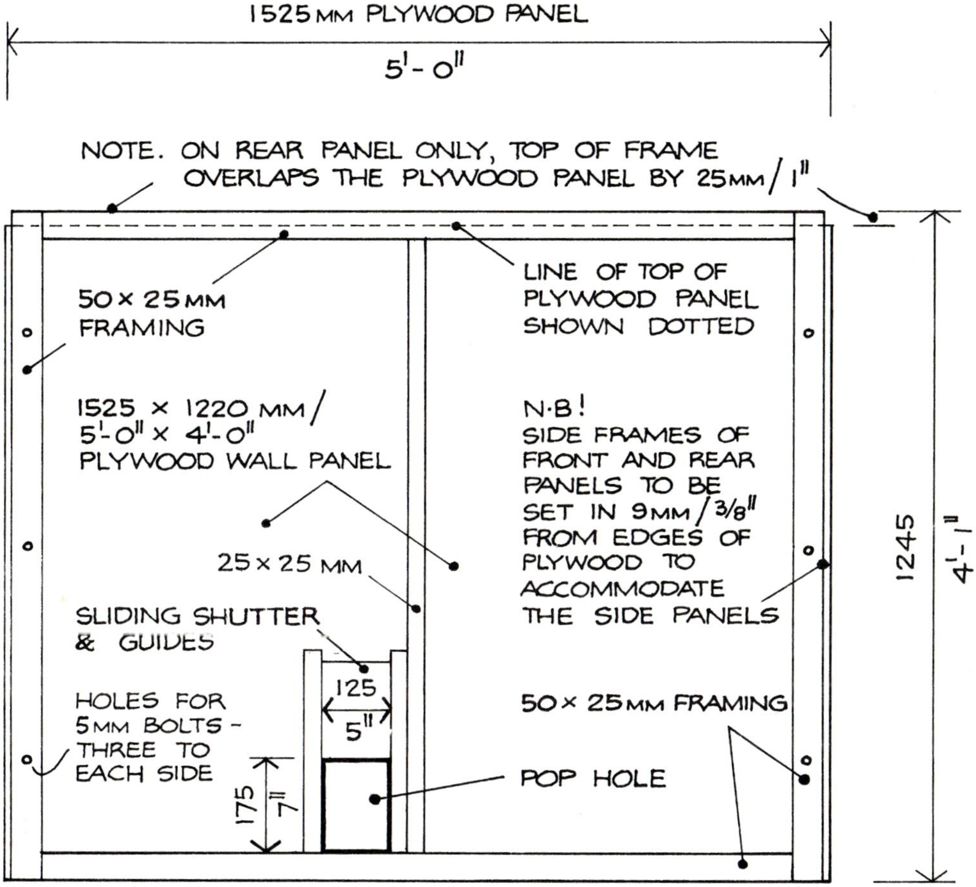

VIEW OF INSIDE OF REAR PANEL, SHOWING FRAME

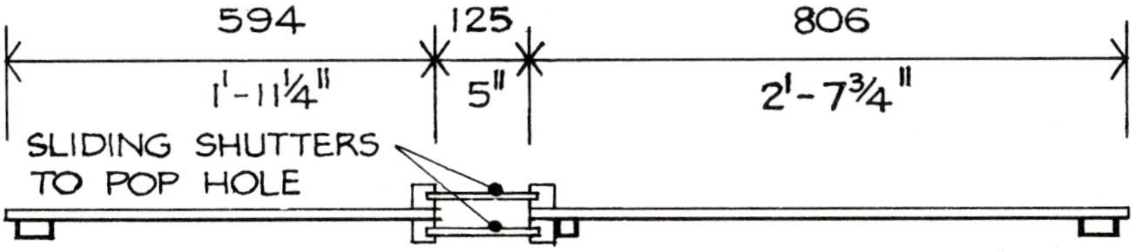

PLAN THROUGH REAR PANEL AT POP HOLE LEVEL

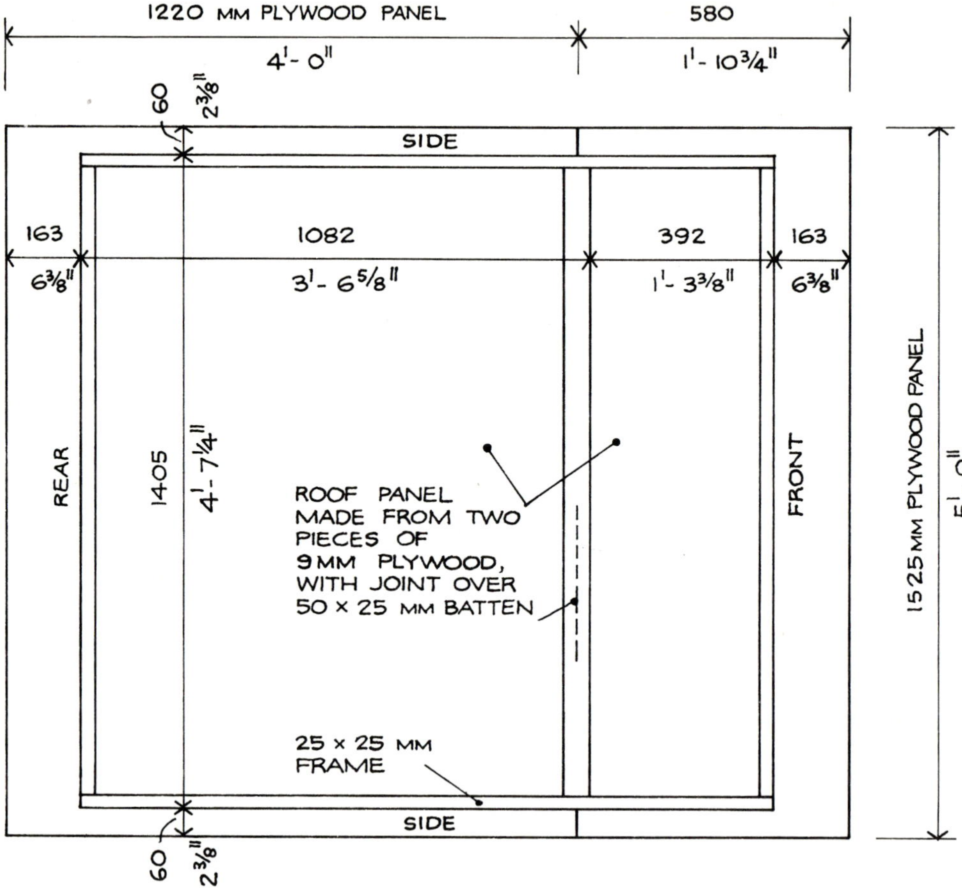

PLAN OF ROOF PANEL, SHOWING FRAMING ON UNDERSIDE

SECTION THROUGH ROOF PANEL

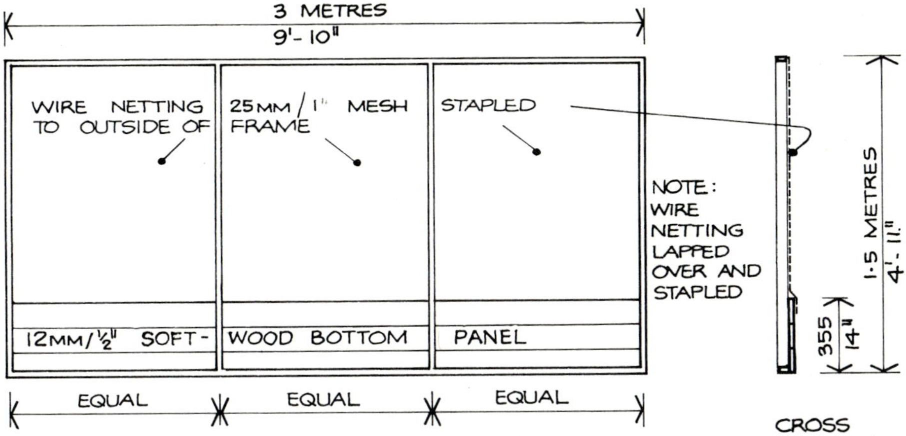

INSIDE VIEW OF BASIC NETTING PANEL

CROSS SECTION

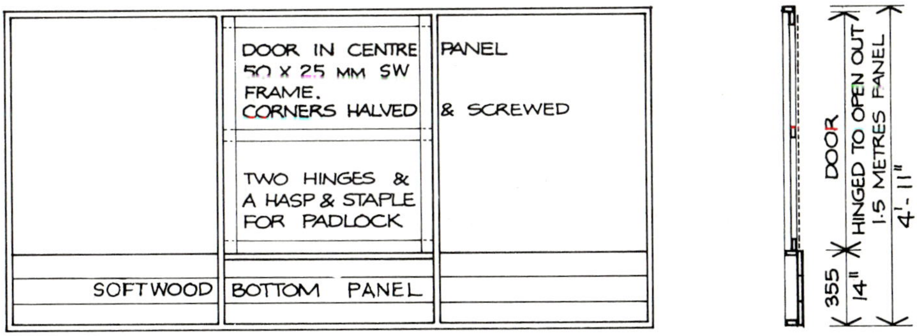

CROSS SECTION

NETTING PANEL WITH DOOR

TRIANGULAR FRAME
TO FORM ENDS OF SUN PARLOUR

FRAME MADE FROM 50 X 25 MM SOFTWOOD,
HALVED AND SCREWED AT INTERSECTIONS.

FRAMES FOR SUN PARLOUR COVERED
WITH WIRE NETTING & POLYTHENE
SHEETING

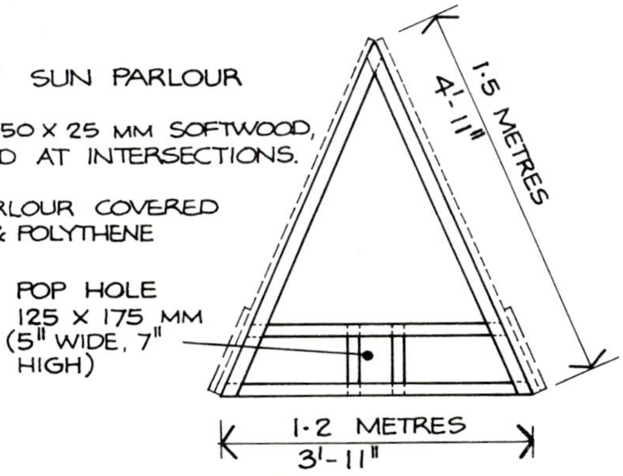

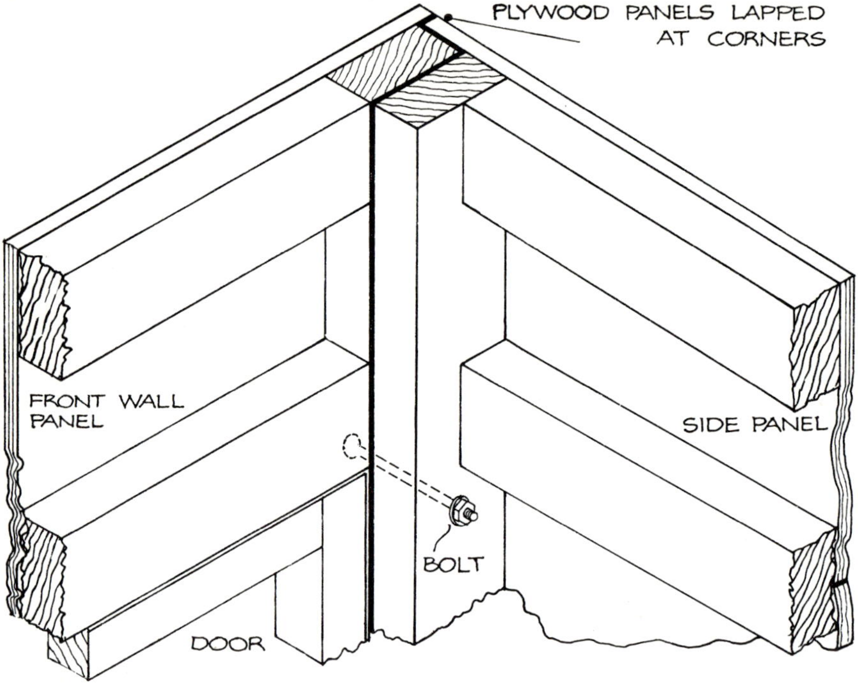

CUT AWAY VIEW OF INSIDE OF CORNER AT TOP OF FRONT AND SIDE WALLS SHOWING HOW THEY FIT TOGETHER

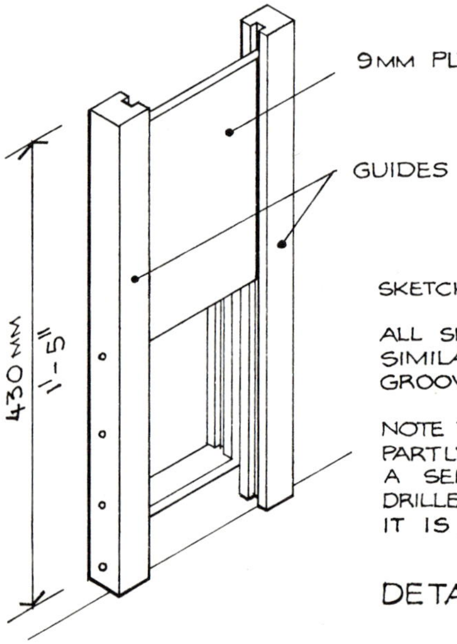

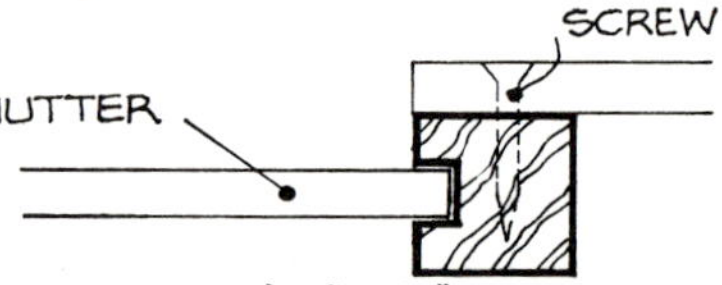

PLAN THROUGH GUIDE

SKETCH SHOWS TYPICAL POP HOLE SHUTTER

ALL SHUTTERS CAN BE MADE IN A SIMILAR MANNER, USING THE SAME GROOVED SECTION FOR THE GUIDES.

NOTE THAT SHUTTERS ARE HELD OPEN OR PARTLY OPEN BY A NAIL INSERTED IN ONE OF A SERIES OF HOLES WHICH SHOULD BE DRILLED IN ONE OF THE GUIDES BEFORE IT IS FIXED

DETAILS OF SHUTTER

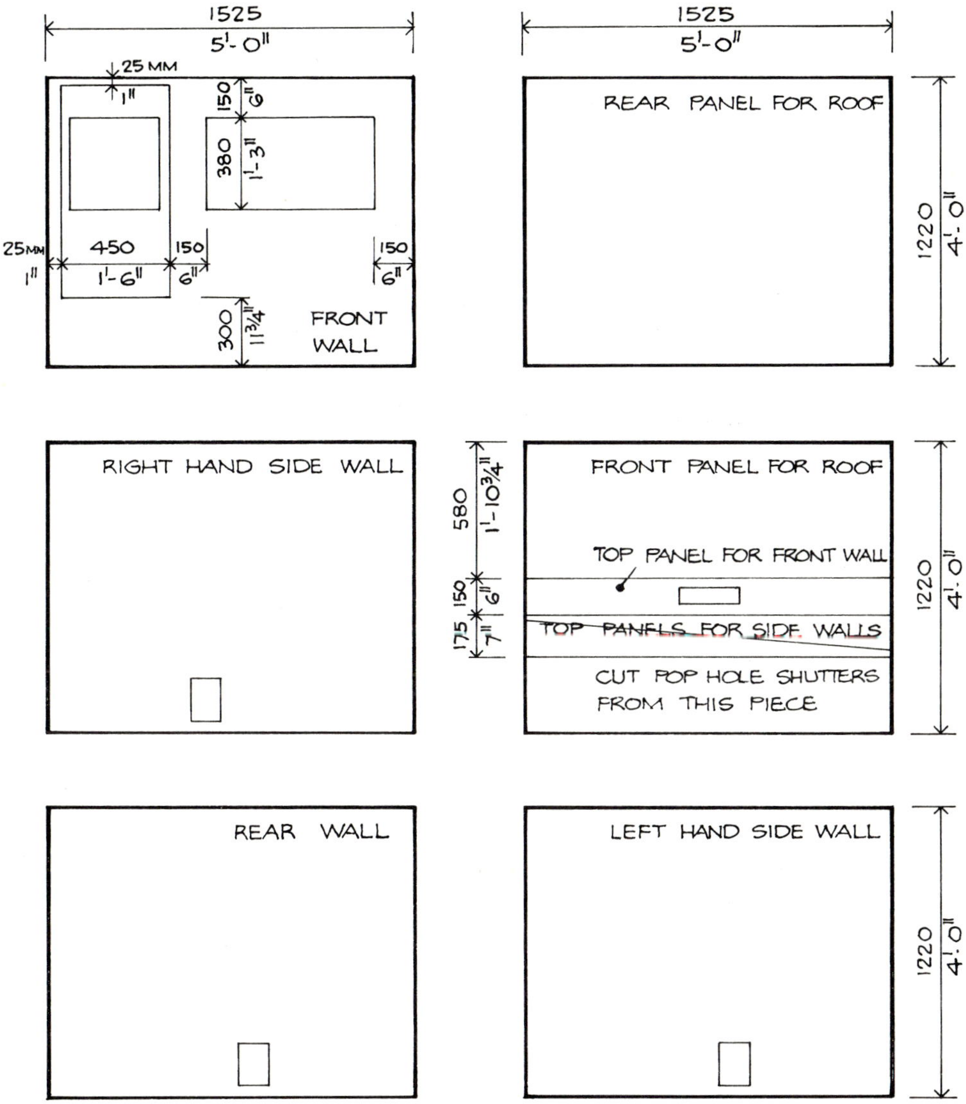

SCHEDULE OF PLYWOOD PANELS

THE ABOVE DRAWING SHOWS HOW ALL OF THE PANELS REQUIRED FOR ONE BROODER HOUSE CAN BE CUT FROM SIX 1525 × 1220 MM / 5'-0" × 4'-0" PIECES OF 9 MM PLYWOOD WITH A MINIMUM OF WASTE.

The economical arrangement of brooder houses and panels

(A) If the brooder house (B.H.) were set up as shown in Sketch A with a pop hole leading to a sun parlour and another giving entry to a pen composed of six panels arranged to enclose an area 6 × 3 metres, we should be able to rear a hundred chicks up to the age of four weeks.

But from the age of four weeks to that of from six to eight weeks, when they will be transferred to the release pens, the chicks need more space – so we add two more panels as shown in the sketch and double the area. After the age of four weeks, the chicks do not need artificial heat and the sun parlour is available to serve as a shelter at night, so the B.H. can house another batch of a hundred chicks if we wish to rear a second lot.

(B) If we leave the first chicks, now four weeks old, where they are but move the B.H. a few feet onto fresh ground and set up another sun parlour with six more netting panels, as shown in Sketch B, we shall be able to rear a second batch of a hundred chicks which will be four weeks younger than the first. When the first batch is transferred to the release pen, we can use some of the panels which enclosed them to enlarge the pen containing the second batch.

Some questions and answers

(Q1) Why are you making a simple matter so complicated?

(A1) I am 'using the plant', and thereby reducing the capital cost of the equipment required to rear any stated number of chicks. The routine described in paragraph (A) is called going 'once through' in the jargon of pheasant rearers. If we add no more than one sun parlour and six panels we double the number of chicks the equipment can rear, and that is called going 'twice through'.

(Q2) Won't bigger roofing nets be needed for the big pens?

(A2) No, all the nets are suitable for pens 6 × 3 metres. When the size of the pen is doubled, you just put two nets side by side and lace them together with string.

(Q3) If reducing capital costs is so important, why not wait until the first batch has gone to the woods and then use the same equipment again without any additions?

(A3) Because saving time is of first importance in the production of top-notch pheasants. A pheasant seventeen weeks old has the physique of a boy of 17 years, and each week or ten days thereafter adds the equivalent of a year to strength and stamina. A boy of 17 may weigh almost as much as he will when he is 25 but the man will outclass the

SKETCH A "Going once through"

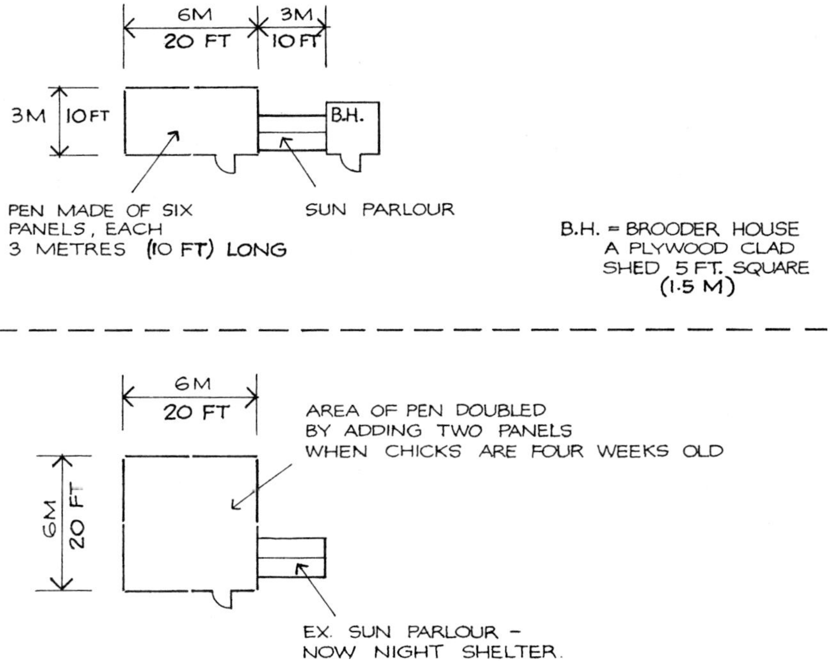

SKETCH B "Going twice through"

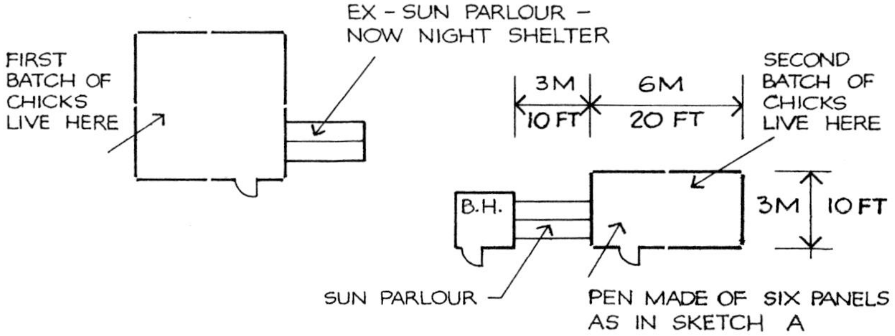

boy for speed, strength and endurance. Eggs set on 15 April will hatch on 9 May so you can work out when the pheasants will be the equivalents of schoolboy 1st XVs, University Blues or the Barbarians; and we are after nothing less than athletes of Olympic standards.

(Q4) I see your point, but if they were not shot until January, surely there could be a third batch, four weeks younger than the second; and then we should have 50% more pheasants from the same equipment. That would be 'using the plant' with a vengeance.

(A4) It can be done; in fact for five consecutive years I did it myself and concluded that it pays only when the third batch is released so that the poults are far enough from their elders to ensure that they never meet. Bullying is the reason for this. Even so, the management of the third batch must be nothing less than excellent and the birds must not be shot before Christmas. When these conditions were fulfilled, the third batch confounded all critics and beggared the pride of some good men. We always shot both cocks and hens hard to avoid producing a strain of pheasants which laid eggs late in the year. All too often, however, standards of management in the rearing field fall and then the third batch consists of miserable little weaklings which are nothing but a disgrace.

(Q5) Well, what's your conclusion? 'Once through', 'twice through', or what?

(A5) A purist would go 'once through'. 'Twice through' is a sound, work-a-day basis; but 'one-and-a-half times through' is better, if you can afford sufficient equipment. Then, if the first batch does badly, you can make up the shortage by rearing a few more in the second.

But if I were a millionaire I should have a lay-out which I have seen but never used. Sketch C shows the scheme. There are more brooder houses and pens than you will ever need, so whether the hens lay well or badly, and no matter what percentage of the eggs hatch, you just put batches of a hundred chicks into each B.H. until you have enough and then release the breeding stock to nest in the wild. To increase the area of the pens you add standard panels to the length.

A little thought shows how easy it is to bring piped water and electricity from the mains to this arrangement; and all the routine chores like feeding can be reduced to a labour-saving drill. Moreover, fewer netting panels are required. One point of detail is worth mentioning: in this lay-out the felt nails which hold the roofing nets should be *on top* of the panels, not at the sides as they usually are.

This is the Rolls Royce arrangement of Fordingbridge-type equipment;

it removes the need for accurate planning and timing and brushes the worry aside with overwhelming weight of plant. The snag is that the amount of capital devoted to equipment might be employed more profitably elsewhere, perhaps in enlarging the shoot.

To some people this constant eye on the main chance is incompatible with sport; they prefer a more carefree approach, even though the inevitable result is a lower level of attainment. Let each choose the path he prefers; I have no wish to make converts, but I enjoy devising schemes which will bring any significant improvement. This extra effort may not be essential if ambition rises no higher than qualified success but it is usually all that stands between the mediocre and the very good.

SKETCH C

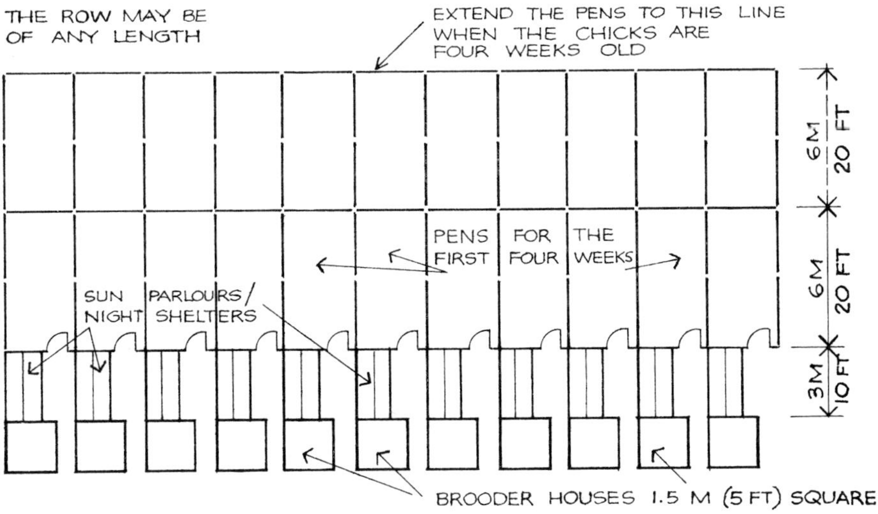

Some golden numbers

If you use Fordingbridge-type equipment and adopt the normal arrangement of brooder houses and pens, the following 'golden numbers' should be remembered:

'Once through' needs one B.H. and ten (10) netting panels, of which two (2) must have doors, for each batch of a hundred chicks. Don't forget that sun parlours are made from standard panels covered with polythene sheeting.

'Twice through' needs one B.H. and eighteen (18) netting panels, of which four (4) must have doors, for two (2) batches each of a hundred chicks.

'Three times through' needs the same equipment as for 'twice through', but rears three (3) batches each of a hundred chicks.

In every case there should be a few extra panels to replace breakages. The man who does not have the brooder house, sun parlours and pens assembled with their roofing nets by 1 April is simply looking for trouble. Many, many novices have left things too late, muddled the golden numbers and have been compelled to call all hands to make netting panels throughout the night while chicks were actually hatching.

Heaters

Where fuel is concerned paraffin is the cheapest, electricity costs about twice as much as paraffin and gas in cylinders (Calor Gas or the equivalent) between three and five times as much as paraffin. Nevertheless, the cost of wages is more important than the cost of fuel, and when convenience and the initial cost of equipment are taken into account, any form of heater may be the best choice for the particular conditions encountered.

A PARAFFIN HEATER The Rupert brooder is heated by paraffin and is designed to house between eighty and a hundred chicks. It can go anywhere – on the lawn, in the orchard, into a disused stable or into a movable pen of any description. This should be remembered because, as long as the release pen is not too far from the rearer's house, a Rupert brooder makes it possible for chicks to be reared within the release pen; and that may be of first importance when it comes to 'releasing'.

The burner must be topped up with paraffin every day and the user must learn how to trim the wicks and clean the flue. It is usually necessary to turn the two wicks up a little at night and down again in the morning, but there is a thermometer.

ELECTRIC HEATERS If there were a reliable supply of electricity from the mains I should choose electric heaters every time. There are three basic types:

(a) The 400 watt, dull-glow coil which screws into a reflector is my, personal, favourite. The Wren heater gives excellent results but the same is true of other makes. The coil should show a thread of light, and not be completely black, so that the chicks can find their way to the warmth.

(b) The 200 watt incandescent bulb, usually coloured red, which fits into a metal reflector and is often used by rearers of pigs. Some rearers of pheasants like them, but I think that they give too much light and that this has undesirable side-effects.

(c) The 'electric hen' which rests on the ground, whereas the two lamps are suspended from the roof. It is the first choice of some experienced rearers.

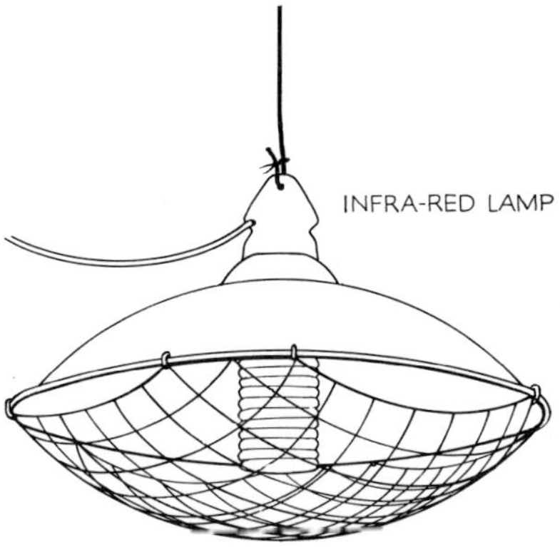

When making a budget, I assume that the heater will be switched on for twenty-eight days for each batch of chicks. A unit of electricity is one thousand watts for one hour, so the cost of the electricity can be calculated.

After the age of four weeks, the chicks should require no artificial heat but, to insure against the occasional frosty night, I run a wire from the brooder house to an ordinary 25 watt bulb in the night shelter. This bulb is screened by a tin to produce twilight.

People tell me that I have a bee in my bonnet about electrical spares because I insist that all the plugs should be identical, and the same with sockets, switches, heating elements, reflectors, fuse boxes and so on. Then anything fits everywhere, and that is a solid comfort when you are making good defects at 3 o'clock in the morning in pouring rain and pyjamas. When preparing for rearing, be careful to undo all the plugs, sockets and switches and clear out the earwigs and insects.

Connecting electric heaters to the mains can be expensive if done by professionals but an ordinary amateur can do it if he uses first-class, new materials and takes pains. Even so, professional advice should be taken at the outset and the finished work should be checked by the same professional.

Heating a variation of the basic rearing equipment by gas.

GAS HEATERS The great advantage of gas heaters is that the same pens and brooder houses can be used whether electricity is available or not. Some of the early models were plagued by minor breakdowns but the new types are trouble free; even so, it is essential that the maker's instructions should be thoroughly understood and followed implicitly.

If Fordingbridge-type brooder houses are used, there can either be one cylinder of gas alongside each brooder or one large cylinder supplying all of them through pipes. In the latter case there will be a stand-by cylinder and a change-over valve, together with accessories such as a regulator and an indicator which shows when the gas is running low.

Any Calor Gas dealer will advise upon the selection of equipment, including the most suitable type of burner, together with its installation.

Setting up Fordingbridge-type equipment

For perfection, the whole of a rearing field – breeding pens, incubator sheds, brooder houses and rearing pens – should be on a well-drained, southward facing slope within sight of the rearer's house. Which is all very nice if you have it, but first-class pheasants have been reared in suburban gardens, so there is a good deal of latitude.

Select a level piece of ground, cut the grass to a height of an inch or less, assemble the brooder house, bring the two panels which will form the sun parlour into position, tie them together with string at the top and erect them snugly against the brooder. Now fit the triangles which form the ends of the sun parlour, making sure that the pop hole is opposite the pop hole in the brooder; then tie the triangles to the panels with tarred string.

Next set up six netting panels to form a pen 6 × 3 metres. These panels should have the wire netting on the inside of the pen so that droppings cannot accumulate on the lower framework. They should be tied together with tarred string no more tightly than will allow them to follow the undulations of the ground.

The next step is to ensure that the lower edges of all this equipment bed closely to the surface so that the chicks are kept in and varmints cannot enter. Then drive felt nails (also called broad-headed nails) $\frac{3}{4}$ inch long into the outer edge of the upper framework of all the panels forming the pen. These felt nails secure the roofing net; they should be about 12 inches apart and should protrude about $\frac{3}{8}$ inch. Once driven, they are left alone throughout the life of the panel. Then fit the roofing net.

If you fear that the whole pen might be blown over, you can put in a few forked sticks as props; they never did me any good, and are a nuisance when the pen must be moved, but they are not actively harmful.

Positioning the heater calls for some explanation. It should **not** be in the middle of the brooder house but nearer to one corner, and its height must be adjusted by pulling or releasing the supporting strings (or chains) *from the outside*. The heater is off-set because we want a gentle change of temperature across the floor of the brooder, so that the chicks can choose the degree of warmth they prefer. The height of the heater should be adjusted, very quietly, from the outside because if you open the door to do so, the chicks may scatter in panic – and fright is very bad for them. More will be said about the correct height of the heater in the section devoted to management but 14–15 inches above the ground will be about right for a trial run with a Wren heater.

If you want a wooden floor in the brooder house, go ahead; I prefer turf with a little pea-sized gravel scattered under the heater. Always make a ramp of earth on either side of the rail crossing the pop holes so that the tiny chicks can scramble over.

The very young chicks do not need the whole of the brooder house for the first few days, and it is essential to exclude draughts, so either put a piece of plywood 18 inches high across the middle from front to back or, better, form a circle with corrugated cardboard as shown in the drawing. This is discarded after about three nights.

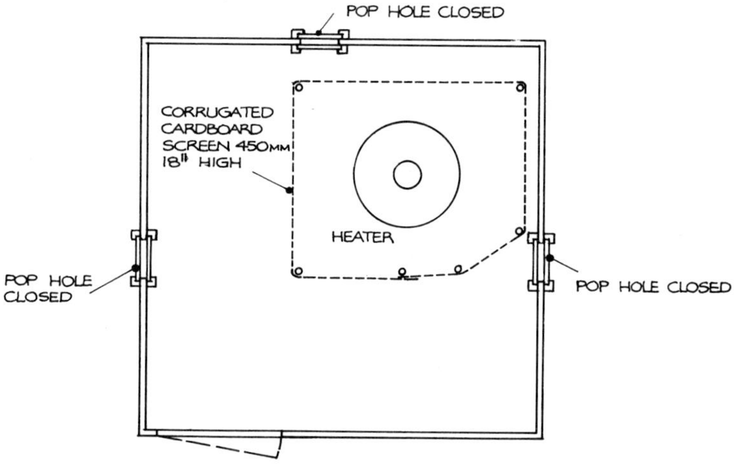

PLAN SHOWING CARD SCREEN CLOSED

An egg tray makes a good feeding trough for the first few days, but the type into which the chicks must thrust their heads is better when they are big enough to use it because the food is less likely to be fouled by droppings. A glance shows whether a drinking fountain made from a glass jar should be refilled; but you may have to put in some pebbles to prevent the chicks getting wet.

The last essential is a piece of plywood 18 inches high held in slides across the lower part of the door; this keeps the chicks in when the door is opened. Toys are not essential but boredom among chicks is not trifling; a bale of straw gives scope for games, and lettuce leaves suspended on strings will be appreciated.

When all is complete, switch on the heater and have a critical look at your work. Make sure that all the doors open the proper way and are on the labour-saving side; ensure that every pop hole has two slides, one on each side – then if you drive the chicks in or out you can keep them where they should be. Is the heater supported by duplicated strings or chains? Many fires have been started when a single support broke. Are there a few bales of straw handy which can be stacked against the outside of the brooder house as insulation if a cold wind rises? Is there any possibility of flooding by heavy rain? If so, either dig some scupper drains or cut enough turf to cover an area 6 × 6 feet and stand the brooder house on it. Only when you are quite satisfied that any fault will have been revealed should you switch off the heater.

Some people deride a smart appearance but I believe that, within reason, a creditable turn-out saves time and improves performance. It is all too easy to allow a rearing field to resemble a shanty town, but then optimum efficiency is impossible and, worse still, pride departs. It costs no more to lay out a rearing field with all the care which would be devoted to a new factory, and that increases the efficiency appreciably; furthermore, men do react to their dress and surroundings and are apt to make a greater effort wherever standards are obviously high in all respects.

Using Fordingbridge-type equipment

If you have never reared pheasants before and are not familiar with Fordingbridge equipment, there is no objection to a training run with day-old domestic chickens, provided that it is held on different ground and the equipment is disinfected afterwards. It must never be forgotten that to mix game and poultry is to invite disease to stay.

Whether you have a trial run with domestic chickens or collect a hundred day-old, pheasant chicks in boxes from a game farm, the drill is the same. If you hatch the eggs yourself pheasant chicks should spend a few hours in cardboard boxes before being treated in the same way as those bought from a game farm. This preliminary period in boxes is described later in this section.

The basic routine is:

(1) Switch on the heater twenty-four hours before the chicks are due, to make everything nice and warm. Adjust the slides which cover the windows so that there is a darkish twilight within the brooder house and close the cardboard circle.

(2) Collect the chicks from the supplier, put them inside the cardboard circle under the heater, leave the brooder house, close the door and watch the chicks through the window for some time.

(3) If the chicks settle down evenly under the heater the lamp is at the correct height and all is well. If they disperse in a ring with a patch of bare turf in the middle, they are too hot and the lamp is too low. If they climb on top of each other under the heater they are cold, so lower the lamp and check for draughts.

Notice that the height of the lamp should be adjusted in accordance with the behaviour of the chicks not, it must be emphasised, to obtain a certain reading on a thermometer. Few novices believe this but it is true.

(4) It may be necessary to adjust the lamp again at sunset, and raise the slides covering the windows if you see fit, but sensible ventilation at all times is essential. Take pains to be inconspicuous and to disturb the chicks as little as possible: no dog should ever be allowed in the vicinity.

(5) Early the next morning uncoil the cardboard as shown in the drawing and allow the chicks into the sun parlour but not into the run. If they crowd together like a rugby scrum in a corner they are feeling cold, so coax them back to the heater. They will soon learn to make their own way to the warmth but, being motherless, are stupid at first.

(6) Enclose the chicks in the cardboard circle for their second night, allow them into the sun parlour as before on the second day and encircle them for a third night unless the weather is distinctly warm.

(7) Thereafter, discard the cardboard, allow the chicks into the sun parlour early every morning and into part of the pen when the sun has dried the grass. Before they are a week old they should be using the whole of the pen but someone must be on hand to drive them into the sun parlour if rain comes.

(8) From the age of one week onwards there is little to worry about. It is better to confine them to the brooder house at night and to deny them the use of the pen in bad weather by day, but a sane balance must be struck between pampering and an over-Spartan upbringing. The commonest faults are confining the chicks in an over-heated and under-ventilated brooder by night and letting them out onto cold, wet grass at dawn; then

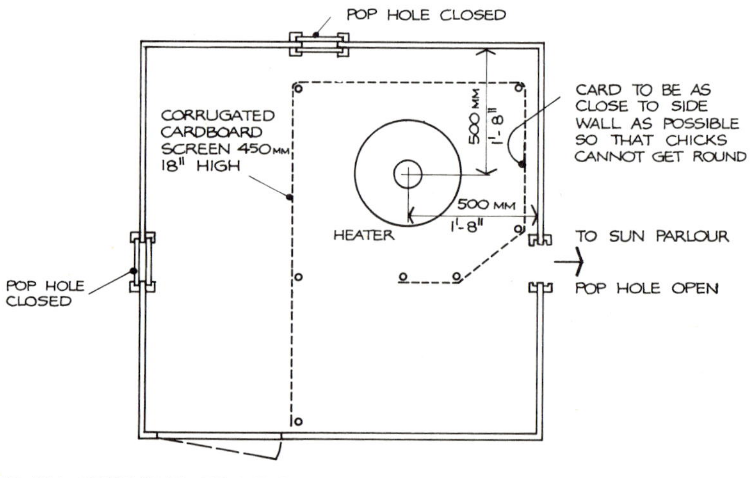

PLAN SHOWING POSITION OF HEATER WITH CARD SCREEN OPEN

they naturally develop chills, cramp and trouble with their lungs. Healthy chicks, like healthy puppies, are quite unmistakable; they are either bright, brisk and energetic or utterly relaxed and asleep; never do they droop.

(9) From the age of two weeks, the heater should be raised about 3 inches and switched off in the middle of the day if the weather is warm. After the chicks are three weeks old, the lamp may be lifted a little more and it should be switched off in the middle of the day unless the weather is particularly cold. The use of the heater should be reduced progressively, so that only on the occasional frosty night is there any artificial heat after the chicks are four weeks old. At that age, the brooder house ceases to be a necessity, the sun parlour becomes their night shelter and the size of the pen is doubled by adding two panels. The birds will live in this larger pen until they are transferred to a release pen when they are about seven weeks old; but the pen should be moved onto fresh ground for the reasons set out in the section on 'Some refinements of the basic method' on page 130.

Preliminary treatment of chicks hatched in your own incubators

Reverting to chicks hatched in incubators at home, it is important that newly-hatched chicks should have a few hours rest after struggling out of their shells; they do not all hatch at exactly the same time, and some are more exhausted than others, so the weak will be trampled underfoot unless they are given time to recover.

Let us suppose that the eggs are hatching on a Thursday and that you have obtained either some of the small-sized cardboard, chick boxes used by game farms, or their equivalent, and have put a little hay in each. Take the dry chicks from the incubator at, say, 9 am and put twenty-five of them into each box. Keep the boxes in a warm room until the evening and then put the chicks under the heater in a brooder house in the manner which has been described already.

By, say, 2 pm some more chicks will be dry, strong and ready for packing in the same way; add them to those in the brooder house after they have spent at least six hours in cardboard boxes. All the foregoing chicks are equivalent to those you might collect from a game farm on a Friday morning.

But what of the chicks which are still hatching, or are weak on their legs, or lack the strength to escape from their shells without help? They should be divided into two parties, and you must harden your heart. Give them plenty of time to dry and gain strength; if they pick up enough to hold their own with the others in the brooder house all is well. If they are only a little frail, they will almost certainly thrive if reared under a broody hen or in a group by themselves; but it is a mistake to persist with any chicks which are deformed, damaged or so weak that they do not pick up within forty-eight hours.

When the cardboard boxes are empty the hay should be burned to avoid infection; I feel safer if the boxes are burned also.

Some refinements of the basic method

The nine numbered paragraphs in which the basic routine was described, contain the bones of the matter, but the chicks will thrive better if some refinements which were omitted for clarity are added; and the chief of these is the constant use of fresh ground. If you are going 'once through' you have one sun parlour and eight netting panels, but the brooder house has three pop holes; so if you shut the chicks in the brooder while you dismantle the pen and parlour and erect them outside a different pop hole, the chicks can go onto fresh ground twice, even if the brooder house is not moved.

If the plan calls for 'twice through' you have two sun parlours and sufficient netting panels to make one pen 6 × 6 metres and another 6 × 3 metres. So if one pen is assembled outside one pop hole and the other outside another, the first batch of chicks can be sent onto fresh ground by opening one pop hole and closing the other.

A little ingenuity will enable you to slide the brooder house onto unfouled ground while the chicks are in a pen. Constant moves onto new ground benefit hygiene in general, and especially reduce the likelihood of coccidiosis, but the chicks also find a wealth of seeds and insects each time they move and searching for them provides an interesting occupation. This is important, because contented chicks thrive where the bored become bloody-minded and develop all the worst characteristics of gangs of street-corner boys. The bullying becomes deplorable and then the lower half of the pecking order most emphatically does not prosper.

I do not remember seeing any serious 'feather picking' when pheasants were habitually reared under broody hens but it is the curse of large, motherless groups. At the first sign of it I fit every bird in the pen with a plastic bit. These are gadgets which pass between the mandibles of the beak, as a horse's bit passes through its mouth, and are kept in place by bending the ends and fitting them into the nostrils. They can be obtained from all suppliers of gamekeepers' stores and are made in two sizes to suit the age of the chicks; the instructions come with the packets. They are not cruel or harmful in any way but they prevent complete closure of the beak, and they stop 'feather picking' entirely. The bits are removed when the birds are put into the release pens.

Poults with plastic bits.

The alternative to plastic bits is trimming the end of the upper mandible either with nail snippers or with a special electrically-heated tool. This is not painful, and the mandible grows again just as our nails do, but I have a slight preference for plastic bits. This trimming of the beak has the hair-raising name of 'de-beaking'.

Every horseman knows that a horse detects fear in its rider at once, and reacts to it. I am convinced that pheasants in pens are conscious of bad temper in a man in much the same way, and that they are frightened of it. Fear is bad for pheasants of all ages; a serious fright stops chicks growing for three days and it reduces the number of eggs laid by mature birds for the same length of time, so a good rearer never startles his birds. He seldom speaks and never moves suddenly, rattles buckets, allows polythene sheeting to flutter in the wind or does anything to attract attention.

As far as 'Fowl Pest', which is also called 'Newcastle Disease' is concerned, the immunisation is simple and cheap, but it cannot be learned from print. If you decide to take this precaution, I suggest that you ask a vet to show you the ropes.

There may be losses of up to 3% in the first three days, which is fairly normal; thereafter, up to the age of release, the losses should not exceed 2%, or 5% at the very worst. If they are greater, look to your management, particularly where chills, cleanliness and diet are concerned.

Feeding chicks

Clean drinking water must be available at all times and it is self-evident that birds in pens must be given a properly balanced diet containing all the essentials. One of the great, labour-saving advances of modern times is that satisfactory foods can be bought ready made, in the form of crumbs or pellets, from any of the major suppliers who advertise in the shooting magazines. Each pellet is said to contain the proper proportion of vitamins, protein and minerals, so the only problem lies in giving the correct amount of the right-sized pellets. The recommended size is printed on the package in some form akin to, 'Pheasant Chick Starter Crumbs. Give these crumbs from hatching to the age of X days.' There will also be, 'Pheasant Growers' Pellets' to be fed between the ages of X and Y days with, possibly, a larger size suitable for chicks between Y and Z days old.

My personal preference is for pheasant foods, although I have had satisfactory results from turkey crumbs. Ordinary domestic chicken crumbs seem to be perfectly all right if pheasant chicks can supplement them with natural food.

Having seen coccidiosis devastate rearing fields I ensure that all crumbs and pellets contain a coccidiostat. When given at a low level, this builds up a resistance to coccidiosis but it does not give complete immunity; nevertheless, it is a safeguard which few experienced rearers ignore. The action to be taken should coccidiosis break out is described in the section on 'Diseases of pheasants' on page 134.

Hand feeding may be necessary for the first few days. Either trickle some food from the fingers or jiggle the trough up and down to tempt shy

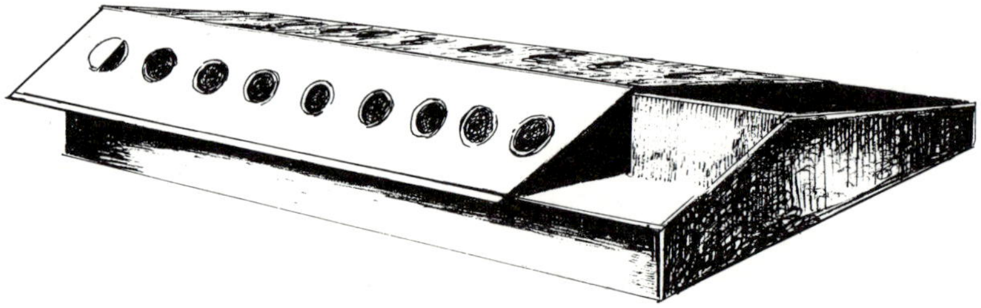

A TROUGH FOR FEEDING CHICKS

feeders, but hoppers or troughs in general are perfectly satisfactory up to the age of four weeks. Thereafter, it is essential to teach the birds to come to the rearer's feeding signal, and whether this be a characteristic whistle or tapping a bucket does not make much difference, provided that it is always the same. The scheme is to put only half a day's food into the hoppers, so the birds will be hungry when the signal is heard and the other half of the ration arrives. They will soon associate the signal with the arrival of food and will come running. The same signal is used to call the birds both in the release pens and, later, to the feeding rides. Hoppers and troughs should always be under shelter within the rearing pens but in fine weather some food should be scattered in the grass to make the chicks work and to give them something to do.

When making a budget or ordering food in advance I assume that a hundred young pheasants will require 140 pounds of chick crumbs and 45 pounds of kibbled (broken) wheat from hatching to the age of six weeks. If kept in pens between the age of six and eight weeks, an additional 105 pounds of Growers' Pellets and 55 pounds of whole wheat will be required.

As a very rough guide to the quantities to be given each day, a hundred chicks will require about 5 pounds of crumbs in their first week; but if you work on the following lines you will soon get the hang of feeding. Measure, don't guess, your estimate of half a day's food and put it into the trough; if none is left when the second feed is due, be a little more generous, but if an appreciable amount is uneaten *discard the old food* and give a trifle less next time.

There is a diagram on page 121 of *The Complete Book of Game Conservation* (Barrie & Jenkins) which shows the average weekly consumption of food by a hundred pheasants from hatching to the age of twenty-one weeks; and I am certain that no one who gave these amounts would be wrong. Nevertheless, I prefer to give as much as the birds will clear up quickly but no more; and the outcome is that my pheasants always seem to eat a trifle more than *The Complete Book* says they should.

Begin by giving the smallest size of crumbs but do not change suddenly to pellets on the appointed day; mix the two sizes together to avoid a jolt. In the same way, mix a little kibbled wheat with the pellets after the age of three weeks and gradually increase the amount until there are two measures of pellets to one of kibbled wheat by the age of five weeks. From six to eight weeks old, pellets and whole wheat should be given in the proportion of two parts by weight of pellets to one of whole wheat.

When doubt arises, remember that too much is better than too little provided that the uneaten food is taken away: to top up a half-empty trough is folly, and bear in mind that pellets are more nutritious than grain. You should learn to manoeuvre troughs and drinking fountains through pop holes, so that they can be changed without disturbing the chicks.

Whenever possible, tie a lettuce, a bunch of young kale or something similar on a string suspended from the roof so that the chicks can fly up and peck pieces off. The traditional tonic for chicks which seem jaded is chopped up hardboiled eggs, and a trace of olive oil with the feed is recommended if the feathers look dry. I ensure that grit of a size suitable for domestic chickens of the same age is always available for pheasants.

Before packaged foods were on the market, young pheasants were always fed on mashes, and every keeper had his own secret recipe. I happen to believe that this is one respect in which the new is a vast improvement upon the old.

Diseases of pheasants

As long as a veterinary surgeon is available to look after the less-common diseases, a novice need only worry about gapes and coccidiosis.

Gapes is caused by little red worms in the windpipe and the disease can be cured but not prevented. The symptoms are a characteristic opening and closing of the beak, a clicking sound and a gasping for breath; losses can be heavy if no remedial action is taken. The treatment is to put one of the new anti-gapeworm drugs into the drinking water in the proportions specified by the manufacturer.

Gapes, like influenza, is always with us but it seems to strike in the release pens rather than in the rearing field. If prompt action is taken there should be few, if any, deaths but if any number of birds in a pen are infected I avoid using that patch of ground for several years. The disease should not kill nowadays but it does check the growth of youngsters.

Coccidiosis is an altogether more deadly menace about which there is a lot of wrong thinking. It thrives in dirt and overcrowding but it can occur no matter what precautions are taken. General cleanliness is the first line of defence, with a coccidiostat in the crumbs and pellets to build up resistance in the growing birds as the second bulwark. Even so, there can be an outbreak if the infection is heavy, and that calls for a calm head and quick action.

A curative dose of some drug must be added to the drinking water, but there is a choice of drugs and, with some, the correct dosage depends upon the amount of coccidiostat in the food. Too much medicine is toxic, too little will not cure; the critical factor is the total amount of the drug given in the food plus that in the medicated water. If you are quite certain that there is no coccidiostat in the food just go ahead and use one of the drugs listed below in accordance with the maker's instructions; but if you have been giving a coccidiostat in the food you should know (a) what drug it contains, (b) at what level and (c) what curative dose of the selected medicine now standing ready and waiting on a shelf in your shed is proper when used in addition to the coccidiostat in the food. And if you have not worked all that out, checked the sums and pinned foolproof instructions onto the wall behind the medicine and measuring glass before 1 April, you have a lot to learn.

A rather out-moded drug with little margin for error in the dosage but still favoured by some who have seen it work wonders in the past is 'Sulphamezathine'. There are several effective alternatives available, among them 'Whitsyn S' and 'Saquadil'. The doses vary between brands and must be measured accurately in accordance with the maker's instructions, but having said that, the most important thing is to act at once.

It should be noticed that the number of days for which medicated water should be given varies both between drugs and with the severity of the infection. Here again the maker's instructions should be followed implicitly – the age-old rural practice of treating any disease with a mixture of all the medicines used previously for different ailments is to be deplored.

Only a pathologist can be certain of identifying coccidiosis, partly because the early symptoms are much the same as those of a chill. If a poult is sitting alone and looking out of sorts, with its feathers fluffed out, its back up, its head sunk onto its shoulders and with no zest for anything, it might have a chill or an upset stomach; but it might have coccidiosis.

Having stopped an outbreak in a pen, or pens, and having established the survivors in quarantine on fresh ground, the question of disinfecting the rearing equipment and the ground itself remains. All equipment can be disinfected with ammonia but nothing can be done about the ground and no one knows how long the disease can survive in the earth; the rearer can only avoid that site. Despite theories to the contrary, lime does not free the soil of coccidiosis.

To emphasise the extent of possible disaster let me tell of two outbreaks which occurred on neighbouring shoots. On one the action described in this section was taken, and three chicks died out of twelve hundred in the rearing field. On the other, the man in charge thought he knew better, and five hundred poults died out of fifteen hundred before a layman intervened and stopped the epidemic within a few days.

There has not been an outbreak of Fowl Pest in this country for some years, so immunisation is less fashionable than it was. It can be done by any layman after the technique has been demonstrated but it is unsafe to rely upon book-knowledge, so those who wish to learn should seek a qualified instructor.

Using the Rupert brooder

A reference to this self-contained, paraffin-heated brooder was made on page 122 when other types of heaters were described. The capacity is from eighty to a hundred chicks; as with all paraffin lamps, the user must learn to trim the wicks, adjust the flame and clean the flue, but that is easily done.

These brooders can be set up almost anywhere, and the pen enclosing the chicks can be improvised from netting of any kind; but let us assume that some spare Fordingbridge-type netting panels are available and one description can then serve for all the possible variations.

All that is necessary is to set up the panels to form a standard pen 6×3 metres, roofed with nylon netting, with a standard sun parlour (made from two more panels) inside it. Then put the brooder close to one end of the sun parlour, confine the chicks to the brooder and parlour for the first few days, but then allow them to use more of the pen as they grow. All the food should be given in the sun parlour and the amount of heat should be reduced in the stages which have been described previously.

The Rupert brooder.

When the chicks are four weeks old, the brooder can be removed for use elsewhere, and then the sun parlour becomes the night shelter, but it is as well to have a small paraffin lamp available to guard against very cold nights. The pen should be enlarged to 6 × 6 metres when the chicks are four weeks old and the feeding and general management is the same as when Fordingbridge equipment is used.

Rearing in large units

This variation will seldom concern beginners but they should have some knowledge of the subject. Broadly speaking, pheasants are reared in large units both to reduce the capital cost of equipment and to enable one man to rear many more birds. It must be appreciated that every increase in the size of the unit demands better management, better hygiene and a nearer approach to perfection all round. The stakes are raised in the sense that one mistake may kill, literally, thousands of birds instead of a single brood; but when success is achieved, the cost of the wages is spread over a far great number of pheasants.

Imagine that there is a disused stable containing fifteen loose boxes. It would not be difficult to put netting over the top of the loose boxes, install half a dozen 'electric hens' in each box and rear several hundred chicks in each of the fifteen boxes. With electricity from the mains, piped water and feeding from hoppers the routine work would be reduced to a minimum. If holes were made through the wall, we could let the chicks into pens of any size we chose.

This is not a pipe-dream; spare buildings are constantly being adapted on these lines to off-set the rising cost of wages, but the management must be nothing short of excellent if all is not to end in disaster.

16
Releasing Pheasants

Releasing young pheasants, that is to say establishing in the woods birds which were brought up in the rearing field, is far more difficult than it was in the old days because we now have to cope with motherless poults. The keeper himself must take the place of the hen in teaching the youngsters how to survive in surroundings which are completely new to them; and no matter how good he is the task calls for every bit of skill he can muster.

It must be admitted that modern equipment has made the rearing of pheasants up to the age of release very easy, but that it is difficult to keep those birds alive and on the property thereafter. Statistics show that only one third of the birds reared are ever recovered; two thirds of these expensive things vanish without leaving a trace and do no good to anyone; if the truth were known, most of the missing were probably dead long before the shooting season opened. It is in this matter of releasing that there is the greatest scope for improvement – if we could even halve the losses there would be twice as much shooting for the same cost, but at the moment it should be accepted that this is a subject about which our knowledge is incomplete. People are reluctant to admit this; the less they know the more do they pound tables, lay down rules which neither work nor stand up to analysis and cook the books to make the results look less calamitous. But the facts cannot be gainsaid.

I can describe the two accepted methods of releasing, and there is also an 'overflowing' method into which we blundered more by good luck than anything else. It doubles the proportion of birds recovered, and is the easiest of them all, but it can only be used in certain circumstances.

Let's get this unorthodox, 'overflowing' method out of the way first. It can only be used when the objective is to establish the poults in the immediate neighbourhood of the rearing field, but for eight consecutive years on one shoot and for five years on another that was the intention, and this is what was done. The last five hundred poults in one place, and two hundred in the other, were never 'released' at all in the accepted sense of the word. They stayed in the rearing field when the others were transferred to the woods but the departure of the majority meant that they could have the use of pens of unprecedented size. These were arranged so that a poult was free when it could fly over wire netting 6 feet high. That was the only way out; escapees could re-enter the pens through 'No Exit' grids if they wished but the only route to freedom was over this high fence. Every year the first poults flew out when they were ten or twelve weeks old and were then eager to rejoin their friends inside, but as muscles and a taste for adventure developed they stayed out longer and went further afield. Nevertheless, they all regarded the rearing field as their home, they habitually roosted in

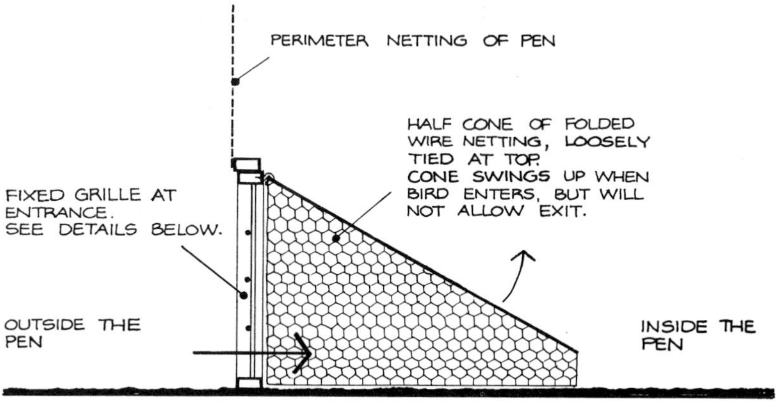

RE-ENTRY BUT NO EXIT GRILLE

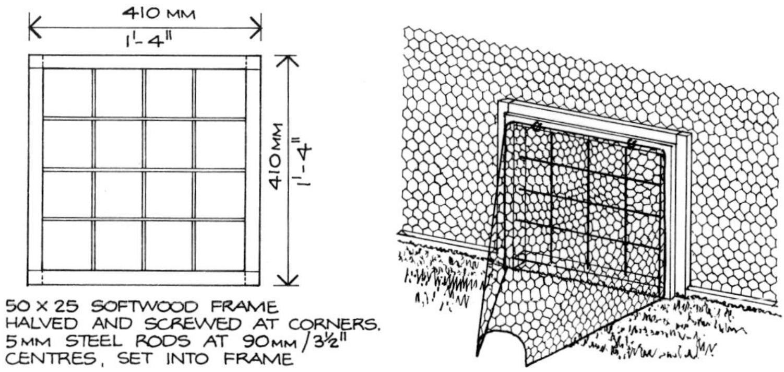

DETAIL OF GRILLE

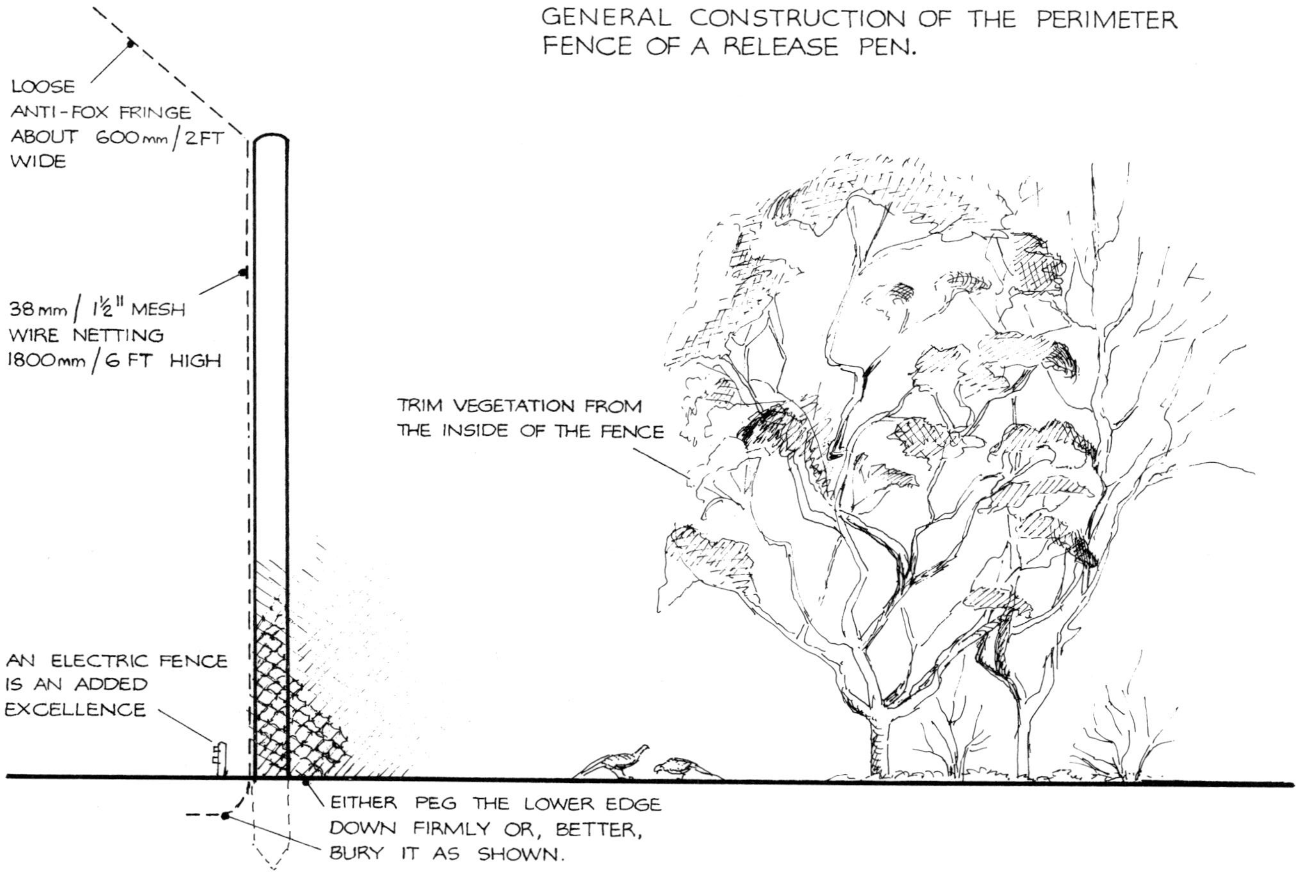
GENERAL CONSTRUCTION OF THE PERIMETER
FENCE OF A RELEASE PEN.
LOOSE
ANTI-FOX FRINGE
ABOUT 600mm/2FT
WIDE
38mm/1½" MESH
WIRE NETTING
1800mm/6 FT HIGH
AN ELECTRIC FENCE
IS AN ADDED
EXCELLENCE
TRIM VEGETATION FROM
THE INSIDE OF THE FENCE
EITHER PEG THE LOWER EDGE
DOWN FIRMLY OR, BETTER,
BURY IT AS SHOWN.

the immediate vicinity and whenever they were alarmed they made for the safe haven of the rearing field as a reflex action.

That pattern of behaviour persisted into January and, as I said, birds released in this manner suffered only half the wastage of the others – a fact which was established by wing-tags. I suspect that it was the gentle introduction to the big, wide world which made the difference. They all severed the apron strings, but in their own good time; there were no shocks, no sudden changes, no terrifying journey ending in a strange jungle and no incomprehensible mazes like fields of ripe wheat where innocent poults might easily get lost and starve to death. Perhaps it should be emphasised in this connection that motherless poults are both incredibly stupid and startlingly bad at self-preservation. Pecking a cat is by no means unusual.

The great, unanswered questions are, 'Should we see the same improved rate of survival if we reared pheasants (perhaps with Rupert brooders or cylinders of gas) within the release pens? And would it pay to do so?' This would certainly be more trouble but if the same amount of money sent more pheasants over the guns the policy would be sound.

The accepted methods of releasing pheasants

The over-all plan is to establish poults which are between the ages of six and eight weeks in large pens set up in the woods, which will become their headquarters. In these pens the young birds will learn to search for food, to roost in trees and to fend for themselves while protected by the keeper. When they are sufficiently strong and wise to face the outside world, they will be allowed to do so but any which seek to return to the security of the pen for a few days will be encouraged to come back. In a sense the release pens are acclimatisation centres between the nursery of the rearing field and the rigours of life in the wild.

An ideal release pen has one third of its area exposed to the sun, one third covered by bushes and one third by trees. It is surrounded by wire netting with a mesh of 1½ inches, 6 feet high. A loose, anti-fox fringe of netting leans outwards from the top of the fence and it is better if there is an electric fence at the bottom. There should be at least one, and preferably four, 'Re-entry But No Exit' grilles at the foot of the fence and the bushes should be cut back from the inside so that poults cannot flutter up the branches and escape by gliding over the top.

There should be feeding hoppers and drinking troughs both inside and immediately outside the pen. I, myself, build a pen of Fordingbridge panels and a night shelter inside a release pen and keep the poults there for two or three days.

If numerous willing hands are available to dig the lower edge of the perimeter fence into the ground as shown in the drawing, this should be done; but never forget that this lower edge must be rolled up when the poults have left and re-buried next spring. Shortage of time has often compelled me to do no more than peg down the lower edge and, so far, I have never had cause to regret it. This is another example of the perpetual struggle between the desirable and the possible, and a rational balance must be struck. It is far better to have a serviceable release pen ready in time than a Perfectionist's Dream only fifteen days late. Debate may be shortened by first studying the Rearing Programme in Chapter 17 and then posing the question: 'If the poults have not gone to the release pen on the specified date, where shall I put the day-old chicks which will hatch on the next day?' The answer is, 'Where indeed?'

It is better to site release pens at the centre of the shoot rather than near the edge, close to a source of drinking water and in a quiet place sheltered from the weather. This is why a clearing within a wood is often the best choice. The oldest poults in any one pen should not be more than two weeks older than the youngest if bullying is to be kept within bounds.

The foregoing procedure is common to all release pens but this is where the two accepted methods take different routes. One school of thought puts a light net over the whole of the release pen and keeps the poults inside until, at the proper time, a small batch (about one in five) are allowed to walk out through an open door. The other has no need of an overhead net because the flight feathers *of one wing only* are clipped with scissors so that the poults cannot fly. They are compelled to stay inside the pen until the feathers grow again and enable them to fly over the perimeter fence.

Both methods have their advocates: a rich purist fits an overhead net because poults with un-clipped wings keep warmer but when expense is important the usual practice is to clip the wings and dispense with the net. If the cost were the same, I should expect better results from a bigger pen without a roofing net.

The management of poults

Always choose a spell of fine weather with warm nights for moving the poults; that is why the Rearing Programme in Chapter 17 allows a margin of two weeks for the transfer from the rearing field to the release pens. On the selected day, remove the plastic bits and clip the wing feathers as you pack them into flat crates or hampers; do not overcrowd them and transport the crates to the release pen with the minimum of delay, noise or fuss. Put the crate on the ground in the release pen, open the pop hole and allow the poults to walk out in their own good time to find a meal ready and

Poults just let out of their travelling crate into a release pen.

waiting. They will probably look jaded for twenty-four hours but will have recovered by the third day.

The usual sequence of events thereafter is much as follows. The poults flourish for two or three weeks, they feather beautifully and are as bonny as could be, but trouble strikes when the feathers on their heads are changing. I suspect that this changing of feathers taxes the physique of the growing birds. The robustness of healthy chicks disappears at this stage and they become more vulnerable in every respect. Trouble often takes the form of feather-picking or the allied vice of pulling out each others' tail feathers, gapes tends to appear at this age and the aggressive birds become especially bloody-minded. The more space there is within the pen, the more variety and escape cover, and the less slum-like the conditions, the less will these troubles be; but there is no sovereign remedy and each man must use his judgment.

The more liberty they have outside the pen the more contented they will be and the better they will thrive but, and this is the vital point, more liberty means greater danger from predators, from exposure, from wandering away and being unable to find their way back and from starvation. There will also be a number of birds at the lower end of the pecking order whose one ambition is to escape from the pen and run as far as possible from those who have bullied them; if they get out they are unlikely to survive for twenty-four hours, so let's deal with them first.

Either wire off one corner of the release pen, or build the equivalent with Fordingbridge panels, and confine the bullied, the under-sized and the feather-picked by themselves. They usually stop feeling sorry for themselves and start growing again within a few days.

If about one fifth of the remainder are allowed to walk quietly out through an opened door, they will probably be delighted with their freedom, peck about in the immediate neighbourhood for a few hours but be only too anxious to return to the main flock in the pen after a time and, if Fate is kind, they will do so through the 'Re-entry But No Exit' grilles. Once the poults are circulating freely from the security of the pen to the outside world and back through the grilles, most of the bloody-mindedness disappears and they flourish again. In this respect it makes no difference whether the birds walk out through opened doors or fly over the perimeter fence.

The difficulty is always to anchor them in the vicinity of the release pen, and a few broods of pheasants reared under domestic hens are valuable allies. The hen in a fox-proof pen anchors her brood, and her poults are a steadying influence on the motherless. Even so, it is the keeper's feeding signal which really holds the flock together and collects the wanderers; and it is useful to know that poults of this age respond to the horn on the Parsons Automatic Feeder. They really do – this is not the wishful thinking

Feeding poults outside a release pen. Notice that the pen has a roofing net and that the keeper is whistling.

of a glib salesman. However, it is on the keeper's judgment and skill that success or failure depends. As time goes on, he will give less food within the release pens and more upon the feeding rides; he knows that pheasants have very accurate clocks in their stomachs, so he arrives at exactly the same time each day, and he patrols the boundary shooing the over-adventurous towards safety.

If chicks are reared under artificial heat for four weeks, sensibly hardened in rearing pens for three more weeks and well managed inside the release pens, the losses should be very small – at any rate until the poults leave the release pen. But it is what happens during the first six or eight weeks after the birds leave the rearing field which is the touchstone of the modern keeper's craft; that is the yardstick by which his skill should be measured. It is probably true to say that at least 85% of pampered chicks which are released badly are dead before they are three months old, and sometimes 95% within two weeks of leaving the rearing field. It is not difficult to keep this wastage within bounds – a novice can do so if he applies himself to the task, but slap-dash methods are doomed to failure.

Some publicity has recently been given to work-sharing schemes by which some members of a syndicate rear chicks at home and hand them over to others, at the age of release, who look after them in the woods. Having seen the vilification which can break out between skilled professionals when this scheme is adopted, I can only say that I vote for each man rearing and releasing the same birds – then no one can steal the glory or escape deserved abuse.

The size of release pens

Let it be said at once that authorities do not agree about the economical size of efficient release pens; there are a number of empirical rules, none of which stand up to a close look. 'As big as possible' is trotted out without the slightest indication of the essential dimensions and questioning seldom produces more than a supercilious silence or raised voices.

The best-known empirical rule is probably, 'one yard of perimeter wire for each poult inside'. Leaving aside the fact that the shape of the pen, whether triangular, round or a long, narrow rectangle, obviously affects the issue, let us concentrate upon square pens. Then if this rule were applied, a hundred poults would have a pen 25 × 25 yards or 6·25 square yards per bird; two hundred poults would have 12·5 square yards per bird, three hundred poults would have 18·75 square yards each and four hundred poults would have no less than 25 square yards apiece. I cannot defend such conclusions on logical grounds and believe that all the empirical rules break down in much the same way when scrutinised.

A self-respecting manager should do better than that because wire netting 6 feet high is expensive and putting it up consumes many man-hours. The one vital question is: 'What is the minimum size of the pen from which a stated number of top-notch pheasants can be released?' I wish I knew the answer, but all I can do is to give the details of sizes which have given excellent results, of qualified success and of outright failure.

The following are the results obtained from some typical release pens of which I have either personal knowledge or reliable, first-hand information. Wire netting was sold in rolls of 50 yards until recently but they are now of 50 metres. All the pens were approximately square.

Pens made of One Roll of Netting. No experience.

Pens made of Two Rolls of Netting. About fifty poults each year did well enough for four years on the same site, then moved onto fresh ground for three years, and then moved back to the original site for four years without trouble.

Pens made of Three Rolls of Netting. Case (a) Five hundred poults were released successfully for six years without changing the ground. Case (b) Five hundred poults developed gapes and coccidiosis during the second year. The site was abandoned and never used again. Cases (c) and (d) Two hundred poults were released for three years without trouble. Case (e) Five hundred poults were released for four years, the incidence of gapes built up progressively but almost disappeared when the pen was moved onto fresh ground, and then built up again. Cases (f) and (g) Five hundred and four hundred poults each year with much the same history as case (e).

Let readers draw their own conclusions from these figures. My own theory is that five hundred birds in a pen made from three rolls of netting are overcrowded, but that this is made just tolerable if the pen is moved onto fresh ground every few years. Conversely, two hundred birds in a pen of that size are in unnecessary luxury.

It is worth noticing that one big release pen is much cheaper than many small pens. If the item in the budget for 'Capital Equipment' had a small surplus at the end of the year, I might well spend it on enlarging the release pens, provided there were already plenty of traps.

17

Producing Pheasants' Eggs

It is possible to breed selected strains of pheasants by penning one cock with five, six or seven hens; and Fordingbridge netting panels built into pens 6 × 6 metres serve well enough if covered with the standard roofing nets. The advantage is that the panels can be used to enclose poults after the laying stock has been released to nest in the wild. The disadvantages are that many small units are much more trouble to look after, and if one cock is impotent the eggs from all his hens will be infertile: a fact which you will not discover until ten days after incubation has begun, at the earliest.

I may be in the minority but I prefer communal pens in which a large number of pheasants are confined with a ratio of one cock to six hens. Such pens should be made of wire netting 6 feet high with 2 inch mesh. A square pen made of two rolls of netting will be 25 × 25 metres and will suffice for fifty pheasants, at a pinch. Three rolls of netting give a pen 37·5 × 37·5 metres which will accommodate 120–140 birds. Even so, rather larger pens, or fewer pheasants inside them, are preferable. Egg-stealing birds, such as rooks, are so common in my district that a light, nylon roofing-net is essential.

The lower 2 feet of the perimeter netting should be covered with sheeting of some kind to cut off draughts at ground level and to give cover from view. Within reason, the more heaps of brushwood, coops, barrels lying on their side and the like which are scattered about the pen the better, because warmth and privacy affect the number of eggs laid. I provide raised rails on which the birds can preen and sun-bathe, together with places where they can dust.

The intention is to procure good pheasants from the woods every year, not to keep the breeding stock in captivity at all times. Furthermore, the breeding stock will be released to nest in the wild as soon as they have laid as many eggs as are required for rearing by hand.

Plans for the rearing season should be complete by New Year's Day and catching the breeding stock may begin in mid-January but must be completed before the end of February. Hazel rod catchers baited with corn are as good as anything for catching and the birds should be hand-picked;

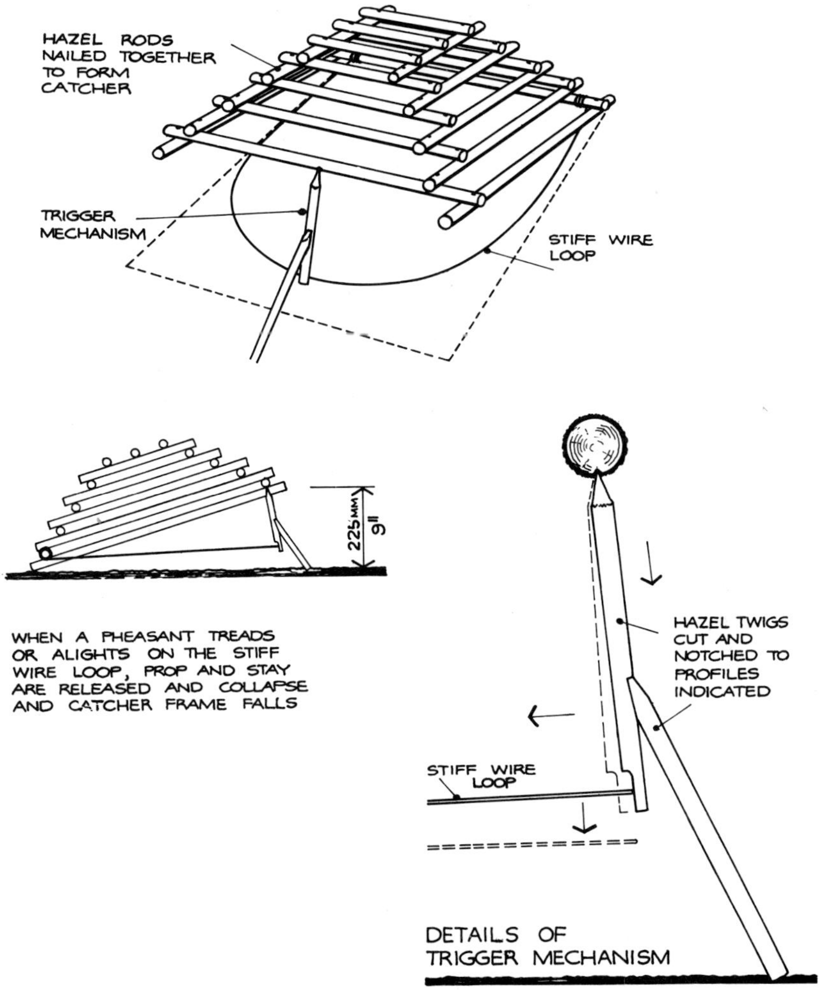

HAZEL ROD
PHEASANT CATCHER

nothing but the best should be kept. Remember that first-season hens tend to lay later in the year than their elders, so mix the ages if you are going 'twice through', and beware of cocks with very long spurs. These may look impressive but they may be past their best for breeding. Put a brail on one wing of all the breeding stock because this reduces the damage when they hurl themselves against wire netting.

Feeding the breeding stock

This is another case of the best being only just good enough. Change the diet gradually from that given on the feeding rides until, by the first week in March, at least 85% consists of first class 'Breeders' Pellets' not, it should be noticed, 'Layers' Pellets'. The balance can be grain, and a little maize is helpful in cold weather; add turnips or cabbage leaves to the diet whenever possible. Ample oyster shell grit should be given together with constant fresh water.

This type of laying ark can be used for pheasants.

How many laying hens do you need?

For planning purposes the critical factors are:

(a) From communal pens at least 95% of the eggs will be fertile.

(b) On average, each hen will lay two eggs every three days but during the first week only half the average number will arrive.

(c) Eggs must be set in incubators within one week of laying.

(d) Well-operated incubators will hatch at least 60% of the fertile eggs set in the course of the season, but individual hatches may be anything from nil to 90%.

(e) The ordinary, 5 feet square brooder house is supposed to accommodate 100–120 chicks but I prefer a maximum of 110.

(f) In practice we always get more eggs and more chicks than the above figures indicates, but they are a margin of safety which can be reared at home or sold.

To digress for a moment: it is quite possible to make a small profit by producing more eggs or hatching more chicks than you need and selling the surplus. Success or failure depends upon the marketing arrangements made before the breeding stock is penned. A standing agreement with a game farm of repute to the effect that they took all surplus day-old chicks at two thirds of the market price was satisfactory to both sides. Attempts to obtain the market price by selling to amateurs are liable to founder, unless it is thoroughly understood that delivery must be made on the contracted day come what may. An amateur gains nothing by insisting that his chicks must be delivered twenty-four hours earlier than agreed (they have not hatched) or by telephoning at the last moment postponing delivery for a week because his brooder houses are not ready (where can you house his chicks?).

To return to the theme of calculating the size of the breeding flock: if you are going 'once through' the only factor is the capacity of the incubators. Subject to that consideration, of course, as soon as enough eggs have been laid the breeding stock should be released to nest in the wild; so catch enough birds to produce the eggs in two or three weeks.

Things are more complicated when going 'twice through' because the hens lay continuously and the first batch must be in the brooder houses for four weeks, the second batch must take their place without delay and the eggs must not be more than one week old when set in the incubator. There is more than one satisfactory solution but the following is a worked-up example. It is the rearing programme of a real-life shoot which set out to rear 1200 chicks going 'twice through'. The available equipment was five

Fordingbridge brooder houses and pens, plus some stables where the chicks could be reared for four weeks before being transferred to pens in the rearing field.

REARING PROGRAMME 1975–6 SEASON

Objective: 1200 chicks
Breeding stock: 72 hens plus 12–15 cocks
Set 170 eggs for batches 1 and 7
Set 333 eggs for all other batches
All surplus day-old chicks to be housed in the stable
Food required by these chicks up to the age of eight weeks: 26 cwt chick crumbs or pellets; 4¾ cwt kibbled wheat; 5¾ cwt whole wheat
Total cost of food: £188·55

Batch	Set eggs	Hatch	Put in brooder	To rearing pens	Release	In which wood
1	14/4	9/5	A	5/6	20/6 – 4/7	W
2	21/4	16/5	B & C	12/6	27/6 – 11/7	W
3	28/4	23/5	D & E	19/6	4/7 – 18/7	X
4	5/5	30/5	Stable	26/6	11/7 – 25/7	X
5	12/5	6/6	A & Stable	4/7	After 18/7	Y
6	19/5	13/6	B & C	11/7	After 25/7	Y
7	26/5	20/6	D	18/7	After 1/8	Z

Batch 7 might be increased or diminished to bring the total to 1200 chicks. Release the breeding stock as soon as the necessary eggs have arrived.

Copies of this programme were given to all concerned. A late spring might cause the whole programme to start one week later, in which case every date moves back one week, but apart from that it should be clearly understood that any departure from the programme will result in additional expense and, almost certainly, in lowering the quality of the birds.

Some professional keepers are able to prepare such a programme but others are not. I feel very strongly that this is a job for the 'Guv'nor'; the manager should take the responsibility for drawing up a programme which can be followed blindly if need be, and if he does not know the ropes well enough to do so, he should set to and learn.

Here is a test question. How many netting panels are essential for the foregoing programme? Remember that chicks in the stable need none, neither do the poults which have been moved to the release pens need any, but those which spent the first four weeks in the stable need big pens while they are in the rearing field.

If you found this problem none too easy, you will probably not delegate the task of drawing up the programme. Neither will you expect a keeper to remember all the dates correctly when distracted by corn merchants delivering the wrong food just as lightning strikes the overhead wires and blows all the fuses. That problem is just one of many which arise in the course of every rearing season, and the 'Guv'nor' should have solved them in advance.

18
Hatching Pheasants' Eggs in Incubators

Until Man intervened, all English birds except the cuckoo built a nest, laid a clutch of eggs, went broody, incubated the eggs until they hatched and then reared the family. This was true of poultry as well as of wild birds so the obvious way to rear pheasants at home was to do a little bird's-nesting, set the eggs under a broody domestic hen and let her bring up the chicks which hatched. When sufficient eggs could not be obtained from the wild, an egg-producing stock of pheasants was caught and penned just as we do today.

Gamekeepers would tour the neighbouring villages buying broody hens in great numbers before establishing each of them in a coop with, or without, a run. A domestic hen can rear from fifteen to eighteen pheasants, so quite a modest shoot would have forty coops, while the sight of several hundred was not rare. The important thing to remember is that each coop contained a domestic hen which had been broody and that every single one needed attention at least twice a day.

The amount of labour involved in any case makes the method too expensive nowadays but the governing factor is that broody hens no longer exist in sufficient numbers. There may well be more chickens in this country than ever before but modern hens are of no use to the gamekeeper because they never go broody. To increase the supply of eggs for human consumption, science has produced breeds which lay and lay without feeling the urge to incubate eggs. The species survives only because Man hatches their eggs in incubators; but for that there would be no second generation of non-broody breeds, so a man who wishes to rear pheasants on any but the smallest scale must also hatch eggs in incubators. And that simple statement heralds more expense and complications than might be suspected.

Basically there are two types of incubators: cabinet machines and those which work on the 'still-air' principle. In one form or another the latter must have been familiar to most of us since childhood. The older models hold 100–120 eggs, are usually heated by a paraffin lamp and are as temperamental as the traditional *prima donna* of fiction. They hatch pheasants' eggs tolerably well if the operator knows how to coax them along (and that needs practice) but each incubator takes twenty-four days to hatch one batch of 120 eggs. If the breeding stock is laying 120 eggs each week, you need four of these incubators and each will work for no more than forty-nine days each year. For the rest of the time the expensive things will stand idle.

The major poultry breeders use big cabinet incubators which are heated by electricity and are fitted with thermostats and devices which automatically control the humidity. They are far more straightforward to operate than the old-fashioned still-airs. You can put in additional eggs as often as you wish until the machine is full, so if the right size is chosen only one cabinet is essential, and they are reliable, in so far as that term is applicable to any electrical appliance. Some users have an automatic alarm and a stand-by generator in case the supply of power from the mains fails.

These cabinet incubators were designed to hatch the eggs of domestic poultry, and they serve that purpose very well, but they are much less satisfactory with the eggs of game birds. All is well for the first twenty-one days but it pays to transfer the eggs on the twenty-second day of incubation to still-air machines and complete the hatching there. There will be notably fewer pheasant chicks if the eggs are left in the cabinet; but fewer still-airs are required because the eggs only spend three days in them, instead of twenty-four.

This combination of cabinet and still-air incubators is the accepted way of hatching pheasants' eggs in bulk; but the rearer who uses it for his own shoot alone must face the fact that it will seldom be in use for more than seven or eight weeks in each year. For months it will be depreciating in value while standing idle – so should he install new machines, which are reliable but expensive, or should he try his hand with cheap, second-hand plant? It is a very difficult decision, but having worked with both I feel competent to express an opinion.

When rearing under broodies was abandoned on the Old Shoot, we bought five still-air incubators of archaic patterns together with two, very old, gigantic cabinets. The second cabinet was a hedge against breakdowns, and was the only precaution we took. The need for safeguards may be estimated from the fact that the whole outfit, including the shed which housed it, cost well under £100. We coaxed those ramshackle, gremlin-haunted antiques into working and we used them for years. Not once did we fail to supply a customer on the agreed date or run short of chicks for ourselves; and the over-all rate of hatching was better than 55%. The capital costs were negligible but the wear and tear on the men was considerable. The need for minor adjustments and repairs was so unending that we lived in an atmosphere of crisis for months every year.

Our target, and we never fell below it, was 300 chicks per week during some years and 500 in others, so the financial stakes were not low and a casual observer might have concluded that we were highly successful. What will never be known is the size of the margin by which we avoided burning the place down, electrocuting all the eggs, chicks and ourselves or falling victims to nervous breakdowns. My point is that battered old incubators can be compelled to serve their turn at no great expense, but that the operators may have a rough passage.

Things were calmer on the New Shoot because a neighbour had bought comparatively trustworthy incubators for his own use and found that he had hatching capacity to spare. Our eggs were hatched by him for an agreed fee and both sides were happy; but the system is not ideal and a central hatching station is preferable.

A central hatching station

The following is a description of the best way of organising the hatching of eggs that I have encountered outside the major game farms. The hatching station more than pays its way and the results please both the owner and many rearers of pheasants in the neighbourhood. There is nothing novel about an owner of incubators hatching eggs for others (many have lost friends, reputation and money thereby) but to make a reasonable profit while earning plaudits from the rearers is so unusual that the system must be worth describing.

The owner's insistence upon 'sell eggs and buy chicks' is one foundation of his success. It is not true that anyone can hand in eggs and receive chicks in due season, but that is the general idea. For reasons which will emerge, only eggs from approved sources enter the hatchery, and their owners agree to take delivery of a specific number of chicks on a stated date – but the rearer sells the eggs to the hatcher and buys the chicks he takes away. This eliminates the type of vilification which surrounds grinding corn for a fee– the farmer always complains that the miller stole some flour while the miller swears that the wheat was poor stuff.

It is also fundamental that the rearer must be content with chicks which were not necessarily hatched from his own eggs. After all, if he contracted to buy 200 chicks after handing in 327 eggs he expects to receive exactly

200 chicks, but it would be extraordinary if neither more nor less than that number hatched. In theory, this could lead to complaints that the chicks received were bred from inferior stock but in practice most rearers are glad to have an infusion of fresh blood.

At the heart of things are two of Western Incubators' 'Turkeybators'. These are cabinet machines fitted with thermostats and automatic humidity controls which take up to 4000 eggs each and run on electricity. On the twenty-second day of incubation the eggs are transferred to still-air incubators which also run on electricity and are made by the same firm. An important feature is that each of these still-air hatchers is designed to cope with the output of one cabinet incubator.

Except that they all work on the still-air principle these modern hatchers have little in common with the temperamental tormentors described earlier. It would be an exaggeration to say that the operator need do no more than make good the connections and turn on the switches but he has a straightforward task with a machine which performs consistently.

Experience shows that up to 85% of the fertile eggs hatch sometimes but that from 65% to 75% is normal. If 60% is assumed as the average rate over the whole season there is a comfortable margin of safety.

One big outfit like this is much cheaper both to install and to run than many small units with the same total capacity. There is a major reduction in the capital required, only a fraction of the labour is needed, and there is also an increase in efficiency. This is because the scale of the undertaking justifies better buildings and vastly more sophisticated methods of controlling temperatures and humidity.

Infection is, however, a major danger and is one reason why eggs are only accepted from rearers of good repute. The policy is to prevent disease rather than to cure it. Before the season begins, all the incubators and the rooms in which they stand are scrubbed and polished until their cleanliness approaches that of an operating theatre; then they are fumigated.

When more is known of fumigating eggs, that may be done also; but at present they are washed before entering the hatchery and, hopefully, this is sufficient to prevent the infection of one unbroken egg by another; so as long as none break inside the cabinets all is well for the first twenty-one days of incubation. On the twenty-second day the eggs are transferred to the still-air hatchers which are housed in a different, sterile room; and after hatching this maternity ward can be disinfected in readiness for the next batch.

Every experienced rearer will see at once that there must be an open-ended element in an organisation of this kind. No one can foretell exactly how many eggs will hatch on any one day, so there must be accommodation for chicks which are not sold immediately as day-olds and also some source from which a shortage can be made good when the contracted sales exceed the number hatched. In this case, the necessary elasticity is provided by the hatcher's own shoot: he produces about 40% of the eggs for hatching and is quite prepared to part with some of his own chicks or to retain and rear any oddments surplus to his customers' needs. Certain brooders and pens are earmarked for this purpose and the total cost of these, together with all the incubators and ancillary equipment, but excluding the buildings, was £3000 in 1974.

The pros and cons

The advantages of this system to the rearers are that they need devote neither capital nor labour to hatching, that they take delivery of exactly the number of chicks they want on a specified date and that they are not compelled to find homes for eggs which are too late for the 'first time through' but too early for the second. Both the hatcher and the rearers benefit from the increased efficiency and lower costs of this relatively big outfit and the hatcher makes enough profit to justify the amount of capital invested. Against this it can be said that the method is open to abuse, smacks of regimentation and makes the breeding of a strain of super-pheasants impossible.

On looking back I do not regret buying those second-hand incubators. They served their purpose and I doubt if pheasants' eggs were ever hatched more cheaply, but I should not choose that course again if I were building up a shoot from scratch. On balance, the best plan might be to buy day-olds from a game farm when rearing for the first time; then to produce your own eggs but have them hatched in a central hatching station until you were well established; and to start a hatching station of your own when you felt capable of filling a role for which there was a demand.

19
Producing Extra-Good Pheasants

Everyone knows that the pheasants of January are so much harder to hit than their cousins which faced the guns in November that they might almost be a different species. Their speed, strength and stamina are of a different order and marksmanship of a different class is required; yet they were all bred from the same stock, fed upon the same food and reared in the same pens. Two months of life, alone, converted them from long handicap players, as it were, into something near scratch.

There are excellent reasons why we do not refrain from shooting pheasants until January, but that is another matter. Nevertheless, the quality of all the birds can be improved by extra-good management and it can be done without reducing the numbers or adding one penny to the costs. November pheasants can become as difficult as those of previous Decembers, or even of bygone Januaries; while the pick of the birds, those which might be regarded as Olympic Games contenders, will be as far above the common run as are Classic horses above selling platers.

The over-all plan is to breed, rear and release the pheasants much as we do now but to ensure that every single thing, from the selection of the breeding stock onwards, is of the highest standard. Then eggs will be produced only by big, strong, healthy birds, which cost no more to feed than do miserable little weaklings, and the same meticulous care will be taken throughout.

An analogy with a racing stable can be drawn. It is accepted that any young horse is likely to improve if it goes into a professional's stable after being trained at home by an amateur. The value of the professional's superior skill is not denied – far from it – but at least some of the improvement is due to better stable management. The professional's horses are never disturbed when they should be resting; his oats and hay may be no more expensive than the amateur's but they are never musty and they are given in the right quantities at precisely the proper time every day. The same theme of higher standards and tighter discipline runs through stabling, bedding, cleanliness and everything else, and the horses react to it. The benefit derived from better management alone must be worth many pounds in a race – and young pheasants respond to the equivalent of good stable management even more than do horses. Not only do they become bigger and stronger than they otherwise would but they take less time in the growing. On any given date they are more mature than birds of the same age which have been 'done' badly.

If that advantage is added to the flying start given by breeding from selected stock the omens are good. It is also vital that the chicks should be hatched as early in the year as may be. Every week is important. The sooner the eggs are put into the incubators the earlier in the season will the birds become formidable opponents more likely to survive their first encounter with the guns and to profit from the experience, both in wisdom and in physical prowess. It is obvious that the chicks reared in the 'second time through' are at a disadvantage of four weeks in this respect, but I shall return to this later.

The scheme in practice

To consider how this plan can be put into effect, let us go right back to the beginning of the rearing cycle and see how the extra-good management can gain a little at every stage on the ordinary. Despite theories to the contrary, I do not believe that any species of pheasant flies better than healthy mongrels do, but I strongly suspect that it is all too easy to produce small, weak strains of any breed, or of mongrels, by second-rate management. And the quickest way to do so is to keep the breeding stock in pens throughout their lives. For a variety of reasons, which need not be set out here, some professional keepers go to great lengths to do this, but I will have none of it. I want nothing but first-class birds caught in the wild as producers of eggs, and I try to exchange a few cocks with like-minded friends to introduce fresh blood every year. As only the best of the birds will be kept as breeders, catching must begin with time to spare and it is evident that they should be fed on nothing but the best.

The chief difference between a good and a bad breeding pen is that the good is amply big enough, clean, green and peaceful. Serenity is all-important in the rearing field. Any wild thing in a cage is prone to panic, and fright or stress undoubtedly affect the physical condition of birds and animals, so an atmosphere of cloistered calm should be maintained. It is a bad sign if the breeding stock slink about and efface themselves, but you are on the right lines if the cocks stand their ground with something about their demeanour which seems to ask, 'And just who may you be, young man?'

While the chicks are in the rearing field there is only a difference in degree between good and bad management. Beyond ensuring that everything is right on the top line nothing can be added at this stage, but it is always possible to improve releasing. Enlarging the release pen by adding one more roll (50 metres) of wire netting to the length of the perimeter fence never did any harm, and sub-dividing the pen into sections which are then used in rotation can be helpful; but a decisive factor is the judgment of the rearer and his willingness to learn from experience. After all, releasing motherless poults in the modern manner has only been widely adopted in the last fifteen or twenty years and the new technique is so different from the old that a lifetime of experience does not necessarily make a man an expert. It is not entirely true to say that both old hands and novices had to learn the new ropes from scratch, but neither is it completely false.

There was no one month when I was aware that I could tell whether the poults in a release pen were thriving, marking time or going back, but there was a time when I knew that I could not do so, another when my theories sometimes proved to be sound and a stage when I would back my judgment with confidence. It is an ability which grows in a receptive mind; and if every detail of the management is known, adverse factors can usually be diagnosed. In real life it is not possible to forestall all trouble; the important thing is to recognise and cure faults before much harm has been done. A special effort should be made to avoid the setbacks which so often occur at the time when the poults are growing new feathers on their heads.

To return to the pheasants which were reared in the 'second time through'. Since they must be four weeks younger than the first lot, we do not hope to find many Olympic candidates among them although, aided by the warmer weather, they tend to gain a little on their elders; but since they sprang from good stock and were 'done' exceptionally well they will be more mature than ordinary birds of the same age. Even so, they cannot be expected to fly strongly at the very beginning of the season, so the brunt of the first major shooting day should be born by the first lot – but not by all of them. The best of the first lot should be released by themselves in isolated coverts, preferably on the crest of a hill, where they will remain almost undisturbed until Christmas. Traditionally, it was said that these birds were being kept in reserve 'until the boys come home for Christmas', but any schoolboy who imagined that he would be anything but a walking gun when those treasured speed-merchants were flushed was misguided. Like wine of a great vintage, they were only shown to those who could appreciate such excellence. As far as difficult pheasants are concerned, those of the first lot which are kept in the background until the end of December should always provide the cream.

Between the first major drives and this super-day it is usual to shoot the pheasants in the order in which they were hatched, and to release them in places chosen with this in mind. It goes without saying that on every occasion they should be presented as well as possible, but there is a hitherto unmentioned ploy, the value of which I have never been able to determine. The scheme is that nothing but cocks should be shot whenever the beaters are going through a covert for the first time in that season; and the pros and cons are worth discussing.

The sequence of thought which favours the plan runs much as follows. Despite all that is said, the first time through any one covert hardly ever comes up to expectations. This is partly because the undergrowth is thick and the beaters have difficulty in flushing the birds; but the root of the trouble is that, on well-preserved land, pheasants seldom use their wings in summertime except to go up to roost. In the autumn they are nothing but a chubby bunch of loafers sadly out of condition, sorely in need of hard physical exercise and training in operational flying. The way to turn them into athletes fit to race for their lives is to accept that the first time through a covert will never amount to more than the second-best, send the beaters through to flush the birds but spare the bulk of them by shooting 'cocks only'. The casualties are no great loss because there are always too many cocks on the ground at the end of the season no matter how hard they are shot.

This argument is not without all merit and it should not be brushed aside as nonsense. It is undoubtedly true that pheasants are easier to flush when the approach of beaters is not a novel experience and that their ability to fly fast and high increases significantly with practice, but the other side of the coin should be examined.

In general, keepers detest any disturbance of the ground which does not yield the maximum amount of sport. Their case could be summed up by, 'What is the use of my toiling to keep the game on the property if you raise Cain without even trying to shoot half of them?' Furthermore, although many experienced managers agree that the second time the coverts are driven usually yield better sport than the first, the question of what is likely to happen later in the season arises. Pheasants are by no means fools and they learn the significance of tapping sticks very quickly. There are always some which lose their heads and fly over the guns but the proportion which avoid doing so by one means or another increases with every encounter with beaters. Indeed, most shoots have stories of individual birds which were constantly seen on the feeding rides for years but which never went over the guns after their first season. Broadly speaking, a pheasant can only be driven a certain number of times so it may be as well not to waste opportunities of shooting it.

The most sensible policy might be to try shooting 'cocks only' on the first occasion any covert is disturbed until the end of November or 7 December, and then to adopt or abandon the scheme as judgment indicates.

20
Accounts and Records

Some shoots prefer not to know how much they are spending; the manager of a particularly happy enterprise had no accounts at all but was almost certain that the overdraft would not amount to much when all the subscriptions had been collected. Even so, I believe that proper accounts are the basis of wise planning and technical improvements. The accounts give the financial news, and the records of bags obtained, the chicks reared and so on show the degree of success attained by certain methods; without this data it is impossible to budget accurately or to plan with certainty. Worse still, without trustworthy accounts and records it is impossible to stand back far enough to see the wood; and a cold, constructive review of the situation should be made every year.

It is true that shooting, like farming, demands a certain continuity of policy, but that does not mean that no intelligence should be allowed to shine through layers of custom; and intelligence must be guided by established facts. Nevertheless, a sane balance must be struck between keeping no books at all and wasting time on trivia, partly because men who are physically tired by working in the open air simply will not tolerate pernickety clerical duties. I suggest that two separate people should be involved; let us call them the 'keeper' and the 'manager' for clarity, although they may well be partners or fellow members. All the serious book work is done by the manager; he pays all the bills, handles the petty cash and keeps an orthodox set of accounts. He cannot do this efficiently unless he, himself, places the orders, so all purchases are made by the manager although the bulk of them will be at the request of the keeper. The only book kept by the keeper is a big desk diary in which he records significant events. These must include the delivery of food and stores, the setting of so many eggs in incubators, the transfer of poults to the release pens, the bags made and so on, each on its appropriate day.

The manager collects the facts contained in this diary and produces useful statistics from them; he also divides the total expenditure into sub-divisions. These must always keep capital costs apart from recurrent expenditure; and fixed charges, such as rent and wages, separate from those which can be varied at will – rearing and forestry are examples of the latter. The rest is fairly obvious; too much sub-division defeats the purpose but there must be enough to relate each major benefit to its cost.

If this work has been done properly, the manager can calculate how many more hens should be penned if a hundred extra reared pheasants are to be shot, how much the additional rearing equipment would cost and how much the bills for food would increase. Much the same applies to every facet of the shoot. If accounts and records are kept properly, on these lines, the manager can plan with certainty and estimate costs with precision; without them he is guessing.

The secretive ways, the caginess, of the old-time keeper were almost a badge of office. If his employer knew of fifty broody hens sitting on eggs it was safe to bet that more were hidden away somewhere so that his hatching rate would appear to shine more brightly than his rival's. It was impossible to tie him down to a firm figure concerning the number of poults released in the woods and nothing would induce him to forecast the merit of the coming season. Some people found this amusing. My own exasperation can be summed up: 'Are we in this together, or are we trying to get the better of each other?' Neither accounts nor records have any value if the 'keeper' and the 'manager' are not frank with each other.

By the same token, any competent man should be able to make a rational forecast; if he cannot do so he ought not to be in charge. Circumstances can change, disaster can take a dozen forms, but it is no disgrace to be proved wrong on an estimate based on the best available evidence.

If the shoot is being run at a canter, as it were, then neither accounts, records of results nor forecasts of the future matter very much; a ready cheque book can solve most problems. But if anything approaching optimum results are the target proper accounts and records are essential and anyone who impedes their preparation is helping the enemy.

In my opinion, the value of a partner or employee is seriously diminished by any lack of frankness, any reluctance to share knowledge or any inability to make an intelligent forecast. In this connection there is the story of two friends who were so depressed by the gloomy forecasts on their home shoot that they went all the way to Denmark for the first, major covert day of the season. Denmark provided a fiasco while the English shoot made a record bag, and who shall say that the recriminations were not fully justified? A manager should do better than that.

21
Organising Shooting Days

The eclipse of the partridge has reduced the effective length of the shooting season to such an extent that it is difficult to arrange a satisfying number of small days without spoiling the great occasions. The snag is that the longer the pheasants are left alone the better they will fly but the less ground there will be for minor days before the first big drives. A man with a dog can spend hours rummaging about in small corners on the boundary, which are beneath the notice of a larger party, without doing any harm; but the real difficulties arise with a syndicate of, say, seven guns. It is difficult to devise satisfactory small days for three or four men without disturbing the main coverts, and it is often impossible to do so for seven guns. So the syndicate must either be divided into groups which have different small days or there must be no shooting before the first covert day.

Purists do not allow the ground to be disturbed by shooting pigeon on the stubbles, although it does no great harm as long as the guns are sensible and some distance from the woods. The same is usually true of a certain amount of duck shooting near the flight ponds. There is also scope for those who enjoy a little shooting combined with a lot of field-craft while 'walking-in' the boundaries. 'Walking-in' is the exhausting, thankless but necessary, task of patrolling the boundary of the shoot and shooing any pheasants which might cross it towards the centre.

A rewarding scheme is for one or two men to stop the end of a hedge while two others walk towards them abreast of a dog which works through the thick stuff. Every pheasant which heads for the centre of the shoot when flushed is given a safe passage; how many of the outward bound draw fire is largely a matter of self-control and conscience. In theory, only old cocks which are leaving the property should be shot.

If the dog will work to hand signals and the men know the ground not one word need be spoken. Such a satisfying amount of field-craft gives the guns and the dogs an airing, accounts for a few predators, causes few casualties to game, disturbs little valuable ground and does something towards keeping the pheasants at home.

A day of driven pheasants

The Director of the Game Conservancy has said that on a perfect day of driven pheasants every gun would fire a hundred cartridges but the bag would only just suffice to provide each with a pair to take home. His point is well made; the objective is a non-competitive trial of skill, not a massacre. But I do not think that the prescription was meant to be taken too literally and I would qualify it on a number of grounds. Any major day is the culmination of a great deal of hard work and it should be taken seriously, savoured and remembered. I want all hands to enjoy themselves and to be proud of their work as they go home; every bird should be picked up and every runner dispatched quickly, but I do not maintain that the pheasants should be so fiendishly difficult that none but experts can hit them. There is a sane balance between cannon fodder and birds which break the hearts of the well-prepared but less talented performers, although the sooner an ill-prepared man sees the folly of his negligence the happier I am.

Although the season opens on 1 October, pheasants cannot be driven properly until the leaves have fallen, so that fixes the earliest possible date. The first time any one covert is beaten out may well be the most productive, although the second often yields better sport. Then there must be an interval of at least two weeks before the beaters go through again and it is possible that there may be a third day of ordinary shooting before the season ends with a few hours of 'cocks only' in January.

Pheasants shot after Christmas are far harder to hit than their cousins which faced the guns in November; but the later the date of the first major day, the greater will be the expense of feeding the birds and the greater the losses from poachers, predators and straying. As a very rough guide the first major day might be in the second half of November, the second on the same ground might be two or three weeks later and there should be one day of 'cocks only' towards the end of January. Whether the number of pheasants available will allow a third day of 'cocks and hens' or of 'cocks only' early in January, is a matter for decision.

The foregoing applies to any one piece of ground disturbed by a day devoted to driving; in the south-east of England it amounts to about 300 acres; so if the shoot covers 600 acres there can be two 'first days', two 'second days' and so on. Each day is made up of about eight drives, five before lunch and three after. The drives should be arranged in a logical sequence so that the uproar of one herds the pheasants into an area which will be beaten out later. A pheasant which has been over the guns once must rest for at least an hour before being flushed again.

The 'keeper' is usually in charge of the beaters and the walking guns while the 'manager' looks after those in front; and a forward gun who cannot touch the stick marking his stand is in the wrong place. It is traditional to draw lots for stands before the first drive and to move up two places on each subsequent drive, but a little intelligent seeding after lunch can add to the general happiness. Lady Luck may send a stream of pheasants all day over one or two men, while someone two stands away scarcely has a shot, so a wise host sometimes intervenes.

The elements of beating are described in my book *Shooting Game*, and I do not wish to repeat them here, but it should be emphasised that there is a vast difference between good and bad driving and the matter repays study. Beaters' dogs, however, are more often used than they were a few years ago and are worth discussing. For perfection, there would be many beaters and no dogs with them, but nowadays beaters are expensive and hard to find so a few good dogs may bring a major improvement.

I wonder how many enemies I shall make when I say that the most useful dogs for beating dense undergrowth, which I have seen, are all Jack Russells. Let me hasten to add that some spaniels are excellent and that other breeds have all sorts of virtues, but Jack Russells are so small that they can penetrate easily where larger dogs cannot go, they have not the strength to do great harm, and if discipline breaks down they cannot run fast enough to avoid capture. A beater's dog must work close to its master and any tendency to riot can ruin a drive; the unhappiest moments of one season were spent counting eighty beautiful pheasants rising from kale and flying over the boundary of the shoot without drawing fire because a beater's dog had gone berserk and could not be suppressed. Even so, a few ordinarily competent dogs with the beaters are usually an asset.

Presenting pheasants well

Much has been written to the effect that we should prefer difficult shooting to large bags, and in general I support that view, but the size of the bag is one important factor and the difficulty of each shot is another. I think that perfection has been reached when a large bag contains only memorable pheasants. Only if a pheasant lives on in the mind was it worth shooting; and that what may be memorable to one man is insignificant to another does not affect the truth. Let each fire or abstain according to his lights.

Although everyone wishes to send nothing but high, fast birds over the guns, this is only possible when the pheasants can be induced to take off from ground which is far above the level of the shooters. This is easily done from the side of a mountain or from certain coverts high up on the Downs, but genuine archangels cannot be shown consistently in any other terrain. Those who shoot on relatively flat ground must make the best use they can of undulations and must seek means other than extreme height to extend the men, and there are some effective alternatives.

Strong, healthy pheasants, the more mature the better, are a fundamental necessity; so is every ploy which will enable them to climb without sapping their stamina, but what makes them relatively easy or difficult is, 'How far in front can the bird be seen? Is it holding a straight course or curling? And can the gun turn round and take a shot behind?'

When the object is to increase the difficulty of the shooting, the man should not be able to fire at a bird which has passed his stand. He should be physically compelled to take every shot either in front or overhead, but he should be given a reasonable field of vision to his front. I maintain that this is provided by a shooting ride whose width is equal to the height of the trees; the bird then comes into sight at an angle of 45° above the horizontal and is safely behind trees when it is a few yards past the gun. The lower the birds the quicker the man must be, and the time available to him can be diminished to any desired extent by making the ride narrower.

Following the same line of thought, there can be alternative positions behind any big tree, each with a tall screen immediately behind the man. The stand closest to the tree is for crackshots, while those further back give progressively easier shooting.

Critics decry these stratagems on the grounds that the essence of shooting high pheasants is the superlative accuracy which is demanded, albeit in comparatively slow time, but that is not a fair cause for complaint. The birds are as high as the nature of the ground permits and some additional difficulties have been introduced to secure an even contest between them and the guns. Admittedly, quick shooting at pheasants of medium height calls for a different skill, but it is not necessarily inferior.

Yet another form of expertise is required if the pheasants are following a curved path, rather than holding a straight course. This baffling flight may be produced either by the conformation of the ground or by a crosswind but it cannot be introduced at will; although it is usually possible to site stands at places where the pheasants habitually change course.

Experience over the years will reveal the most effective way of organising the drives but I feel strongly that placing the stands on each and every drive is an important detail which should be decided by the keeper and the manager in solemn conclave. All too often there is a lack of imagination; the guns are in a rigidly straight line at intervals of 50 yards,

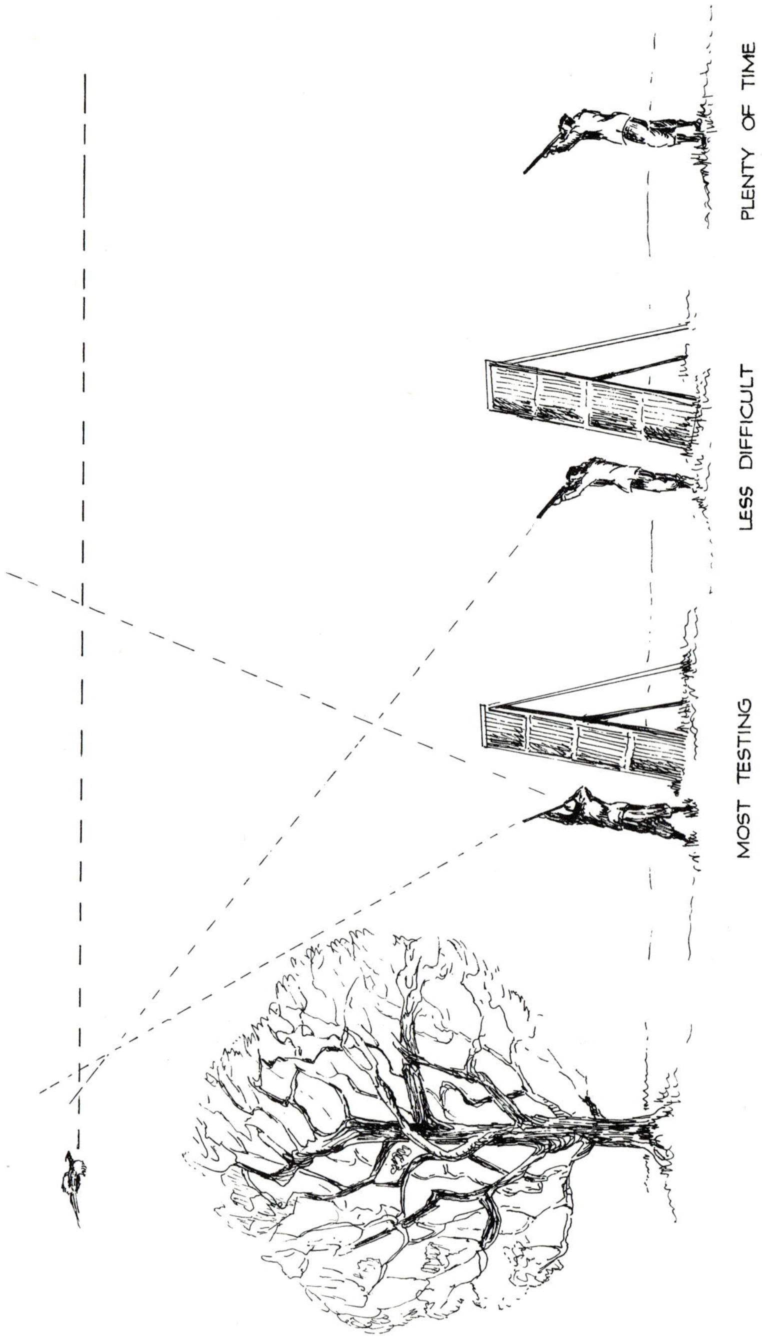

SITING STANDS TO MATCH THE SKILL OF THE SHOOTER

with the foreseeable result that the three guns in the centre have all the shooting while those on the wings have none. Would not all hands enjoy the drive more if the stands were 30 yards apart and anyone who shot his neighbour's bird got an almighty rocket? In the same way, when experience has shown that the men on the extreme flanks can expect little, a second line should be considered. Are the birds so few and the men so expert that a couple of longstops will not be brought into action?

All shooting men should know that cattle have an urge to knock down the sticks which mark the stands and to eat the numbered cards; the arrangement which was a model of clarity on Friday evening may be highly misleading by Saturday morning and guns should be prepared to search for fallen sticks and to deduce their numbers.

The great test of the ability of both keeper and manager comes late in December when the pick of the guns are pitted against the best of the pheasants. They will be acutely conscious of the humble origins of the birds, as well-known shots with 'best' guns and specially-loaded cartridges assemble. They will reflect, rather bitterly, that the resources of the entire world are massed against their homely improvisations. The produce of second-hand incubators and home-made brooder houses, bred from strays and reared in the orchard, can stand no chance against skilled men armed with the pride of the gunmakers' workshops. The treasured pheasants must face the strongest opposition which science can devise and craftsmen can build, and ignominious massacre is inevitable.

Hope will flicker as their pheasants take off like fighter aeroplanes, fly level to gain speed but rise again as they catch sight of the guns. If they have the strength to increase their pace while climbing they still have a chance; and when they stream safely through the barrage few men will feel more pride. All the toil is forgotten as their pheasants – the birds they bred and nurtured – beat the best that Purdey and Holland, Boss, Eley and all the shooting schools can bring against them. There and then it will be decided that those pheasants, and those alone, will be the parents of the next generation.

22
Everyday Problems

Where to seek advice

The best advice on all technical matters pertaining to game comes from the Game Conservancy. This organisation has its headquarters at Burgate Manor, Fordingbridge, Hampshire (Tel: Fordingbridge 52381), but routine problems can be solved by reference either to *The Complete Book of Game Conservation* (Barrie & Jenkins), which was edited by Charles Coles, the Director of the Game Conservancy, or to one of the Green Booklets. These Green Booklets each deal with a specialised subject such as *Rabbit Control, Forestry and Pheasants* or *The Feeding and Management of Game in Winter*; but the best starting point for a novice is probably *Game on the Farm*, which is a digest of several booklets. They can all be obtained from the Game Conservancy.

For advice on major matters of policy such as, 'Should I concentrate on pheasants or on partridges?' or, 'Is this land beyond redemption?' I should seek the help of the Game Conservancy's Advisory Service. In 1976 the charges for a visit by a member of the staff were £40 for a full day, or £28 for a half day for members (£60 a day for non-members) plus VAT. This is a considerable sum but savings can be made if two or three small shoots share the cost. A material consideration is that £40 would be recovered by saving 13 hundredweight of the corn put down as food or by adding forty-five pheasants, or no great number of duck, to the bag – and those modest targets might well be achieved very quickly, leaving the future as clear profit.

The annual subscription (1977) to the Game Conservancy is £15, or £10 a year if convenanted over seven years, and it is hard to believe that any amateur gamekeeper or manager of shooting could make a better investment. I know of no other well-informed source from which disinterested advice can be obtained.

There is one exception to the foregoing, because I turn to any of the old-style, professional keepers who served a proper apprenticeship for guidance on practical trapping. With traps and snares these 'professors' are in a class of their own and unbeatable; I have yet to meet an amateur who was even in the same league but, and I may make a host of enemies by saying this, not all of them have kept abreast of recent technical developments in other fields.

Poachers

There is less poaching than there was because there is less poverty. If you have the ill-fortune to be raided by an organised gang bent upon large-scale looting, I suggest that you should inform the Police and counter-attack with vigour. Your objective should be heavy fines with all the guns confiscated and the cars impounded – and then tell WAGBI (Wildfowlers' Association of Great Britain and Ireland) for good measure. But it seldom pays to prosecute unless you are prepared to go to these lengths.

It is more effective, and usually easier, to turn a minor poacher into an ally by removing his reason for poaching. If he is hungry, find him a job; if he does it for fun, engage him as a part-time trapper who draws his pay in pigeon shooting, with the proviso that he must telephone before bringing his gun onto your land. And if he comes onto your shoot because you have ample pigeon when he has none, try inviting him to shoot with you. With any luck his sense of decency will stop him poaching again.

Almost nothing passes unseen in an agricultural community; early risers and late goers-home, postmen and men working in the fields maintain a watch which ensures that no habitual poacher can remain unsuspected. The one thing which enables him to survive is the Mafia-like silence which prevails if he has the sympathy of the neighbourhood. If you can remove that veil, strip away the false image of a gallant little fellow outwitting a tyrannical foreigner, and get the watchers onto your side, the local poacher will be helpless. For if he sets one foot on your shoot the grapevine will tell you.

Miscellaneous problems

These are so diverse that an imaginary conversation may provide the best thread of continuity.

(Q) When my neighbour is shooting, a lot of his pheasants come onto my ground. May I stand by the boundary and shoot them as they fly in?

(A) In law you may, but there are few quicker routes to unpopularity. Game belongs to the man who owns the land below it, so when his pheasants cross the boundary they become yours, even though he bred, hatched, reared and fed them.

(Q) Whatever the law may say those pheasants are morally his. Is he allowed to come onto my place and chase them back?

(A) He has no right to come without permission; neither may he send a dog to push the pheasants where he wants them.

(Q) But that's what he always does. Can't I shoot the dog?

(A) No, game counts as wildlife. If the dog were chasing chickens or domestic animals it would be committing an offence, so you might have a case if your pheasants were confined to a pen. In the bad old days, this legal point resulted in dogs being poisoned.

(Q) I was enjoying what little shooting I have when the whole of the local hunt, hounds, horses and followers on foot came rushing through and scared every bird out of the parish. Must I endure such treatment?

(A) I sympathise, the same fate has overtaken me; but a letter to the MFH will ensure that it does not happen again. The best plan is to let him know, in writing, at least a month in advance of any time at which you want the shoot to be undisturbed.

Of course foxhounds must be kept away from poults but they have surprisingly little effect upon wild pheasants when they are mature; even if they scatter, they will almost certainly be back within two days. Most Masters are reasonable people but if it comes to a crunch any landowner can warn the hunt off his land; whether the shooting tenant can do so depends upon the terms of the lease.

(Q) Is it true that shooting and hunting can live happily together? That foxes do not harm game?

(A) Let's take these questions separately. Shooting and hunting get along amicably if the people concerned wish to do so. It may or may not be true that the hunting would be better if there were no shooting, but it cannot be denied that foxes are among the great predators of game: if there were no foxes there would be more game to shoot. I happen to enjoy both sports, and even if I did not I should be prepared to sacrifice a little shooting so that the hounds could have enough to hunt, but let there be no hypocritical nonsense about foxes not harming game.

If all hands accept that the presence of even one fox automatically means less game, and is well worth the deprivation, I favour peaceful co-existence; but shooting men who spare foxes should be given the credit for their unselfishness, or why spare them?

(Q) That's fair enough; but what about Gundog Trials? As all field-sports are under pressure, isn't it our bounden duty to support the others and let the trials people have their fun when we are shooting anyway?

(A) You may be right, but I have troubles enough already. I did not go to the toil and expense of building up this shoot from scratch for the benefit of anyone who, now that things are going well, wants something for nothing. I do not ask for money but let them do something useful. Let them hunt round and find a farmer or corn merchant who wants to sell a few tons of low-quality wheat cheaply, or something like that, and I'll pay for it willingly. Then I might feel better disposed; in fact I should be well disposed.

23

Before Parting

Game is a crop, like any other, and the primary reason for shooting is that it provides the best and the most humane method of gathering an important amount of excellent food. It is regrettable that if man is to eat meat creatures must die, but once that has been accepted it is difficult to find any logical reason for excluding enjoyment from the necessary task of filling our larders. There are some whose pleasure is confined to the harvesting; marksmanship, of which the interest lies in the problem of doing something difficult as well as one can, is the be-all and end-all of shooting for them, apart from their contribution to the kitchen, and their pleasure ends when they rise from the table. But they are missing the greater part. Just as a spectator with some training in any art – be it music, acting, the ballet or any other – observes a performance more acutely and appreciates it more deeply because his perceptions have been sharpened, so can a shooting man increase his enjoyment by widening his knowledge of the subject.

There is more craft in beating than meets the eye; driving partridges is a skilled business and the training and handling of gundogs has a fascination of its own, but to those with a taste for such things building up a shoot is the most rewarding of all. Indeed, the primary object often becomes submerged and almost forgotten. As soon as a wish to improve a shoot grips the mind it provides an abiding interest not only throughout the year but from childhood to old age, and the end of learning is never reached. When some ploy succeeds, the lessons are absorbed and when things go wrong means are devised which will serve better next time. The greatest satisfaction, however, comes from watching the progress over the years; the journey is much more important than the destination.

That your shoot, in your lifetime, will never be more than a pygmy when compared with the giants of King Edward VII's day, does not make an atom of difference; in your eyes it is a youngster with sundry shortcomings but developing nicely. That is all you can hope for, but every facet of life becomes more interesting. No matter what you are doing, whether it be travelling by train, sitting by the fire or waiting in a traffic jam there is always something enthralling to think about. If you are press-ganged into visiting Kew Gardens, for instance, boredom may be kept at bay by exploring the possibilities of improving the coverts. Indeed, shooting men often wonder how non-shooters endure the tedium of much of modern life. In the absence of shooting how can they possibly find enough to occupy their minds?

I confess that I have never been able to frame an adequate answer to the question, 'And what have you to show for twenty years' work except a tatty rearing field and innumerable fired cartridges?' I freely admit that there is little enough which would appear in a balance sheet and that there would be more in the bank now if the same effort had been devoted to making money, but the man who asks such a question is dull of soul. The proper riposte is, of course, 'And what have you to show for all your wealth except a lifetime spent running in blinkers?' But that does no good because even if the pompous ass knows what blinkers are, and how much they restrict the view, he probably likes those he wears. He cannot possibly understand the pleasure which comes from seeing a brood of wild ducklings on the pond, from the whistle of wings in the night which tells of mallard, or from counting the number of pheasants which crow as they go up to roost. Because his unceasing devotion to the main chance is fuelled by the urge to pile triumph upon success, he may glimpse the satisfaction which comes from seeing the shoot improve from year to year, but he has probably closed his mind to the possibility of finding any virtue in home-made shooting.

In reality, the financial rewards are not inconsiderable. The record bag on both the Old Shoot and the New is about 120 driven pheasants in one day. That is a trifle when compared with some bags, and there was a certain amount of paid assistance by the time that level was reached, but many syndicates whose six or eight members each subscribe the cost of more than 2500 cartridges every year do worse. When one considers that this measure of success was largely achieved by impecunious amateurs starting from scratch, it would be churlish to deny that they had some success.

The casual observer might well be misled by the brave front presented on both these shoots. It is conceivable that he might believe himself to be watching the idle rich enjoying themselves expensively; seldom would he suspect that the men who are shooting part with less cash for their sport than do many football fans. They also pay less than do many who shoot nothing but clays and the cost is below that of all but the cheapest holidays in the sun. The figures would be very different if the amateurs were paid for the work they do, but if that were the arrangement, the shoot would never have taken shape. The land would still be given over to predators and almost devoid of game, just as it was before the amateurs set out to make what they could not afford to buy. It is something of a paradox that most of the effective conservation of wildlife is done by shooting men and that much of their enjoyment comes at times when they have no gun in their hands.

Count Louis Karolyi showed the heights to which orthodox methods can carry persistent men. Whether you find those peaks intimidating or inspiring may depend upon your point of view, but it should never be forgotten that all progress depends upon doing the work in the one profitable order. If this sequence is followed, it is almost impossible to avoid making some advance, but to depart from it is to waste all effort. The signpost on the path to Count Louis' Nirvana reads:

1 Exclude trespassers.
2 Harry predators.
3 Ensure that game finds food.
4 Provide a congenial habitat.
5 Rear by hand if so minded.

Appendix A

Some useful addresses

The Game Conservancy
Burgate Manor
Fordingbridge, Hampshire

FOR THE PARSONS AUTOMATIC FEEDER
E. Parsons & Sons Ltd.
Dept. S/1
Blackfriars Road
Nailsea, Bristol

FOR FENN TRAPS
A. Fenn
F.H.T. Works
High Street
Astwood Bank
Redditch, Worcestershire

FOR ELECTRIC HEATERS
Diamond Edge Ltd.
126 Gloucester Road
Brighton, Sussex

FOR GAMEKEEPERS' SUPPLIES

Gilbertson & Page Ltd.
Hertford

Ormsby Game Services Ltd.
North Ormsby
Louth, Lincolnshire

Spratts' Game Foods
Evans Mill
Magna Road
South Wigston, Leicestershire

Carta Carna Game Foods
F. C. Lowe & Son Ltd.
Carta Carna Works
Sittingbourne, Kent

The Wildfowlers' Association of Great Britain and Ireland
Marford Mill
Rossett, Clwyd LL12 0HL

Appendix B

Further reading on allied subjects

The Complete Book of Game Conservation ed. Charles Coles, Director of The Game Conservancy (Barrie and Jenkins)

Pigeon Shooting Archie Coats (André Deutsch)

The Young Shot N. M. Sedgwick (A & C Black)

Shooting Game Michael Kemp (A & C Black)

Shotguns and Cartridges Gough Thomas (A & C Black)

All the Green Booklets published by The Game Conservancy

Index